Biological Nitrogen Removal Activated Sludge Process in Warm Climates

*Full-scale process investigation,
scaled-down laboratory experimentation and
mathematical modeling*

Cao Ye Shi, Wah Yuen Long,
Ang Chee Meng and Kandiah S. Raajeevan

Publishing

Published by IWA Publishing, Alliance House, 12 Caxton Street, London SW1H 0QS, UK

Telephone: +44 (0) 20 7654 5500; Fax: +44 (0) 20 7654 5555; Email: publications@iwap.co.uk
Web: www.iwapublishing.com

First published 2008
© 2008 IWA Publishing

Printed by Lightning Source
Cover design by www.designforpublishing.co.uk
Index prepared by Dr Susan Boobis

British Library Cataloguing in Publication Data
A CIP catalogue record for this book is available from the British Library

Library of Congress Cataloging- in-Publication Data
A catalog record for this book is available from the Library of Congress

ISBN: 1843391872
ISBN13: 9781843391876

Cover Photographs

Front: Bird-eye view of Bedok WRP. The effective volume of the activated sludge tanks of each individual train is 8 724 m^3.

Back: Laboratory-scale activated sludge system with an effective volume of 5.6 litres of the activated sludge tanks only, which was adopted in this study to simulate the performance of the full-scale activated sludge process.

Books are to be returned on or before
the last date below.

Bio   vated
Slu   s

LIBREX–

Contents

Contents   vii

Preface

Singapore, as an island country, enjoys a 'garden city' reputation in the world for its beauty and clean environment. However, Singapore has been facing serious water scarcity although it owns a first-class water supply and sanitation infrastructure. Tremendous efforts, both physical and human, have been invested in the recent years to achieve stable and sustainable water supply. The NEWater programme, which purifies used water for potable and non-potable reuse, provides a new water tap and opens a new chapter in water reuse. To cope with the feed water quality requirements of NEWater production, parts of the activated sludge processes in Singapore were upgraded mainly to incorporate biological nitrogen removal. This led to the project entitled "Application of Pre-Denitrification Technology in Bedok, Kranji, and Seletar Water Reclamation Plants" (Contract Ref No: SUI/2001/030), scheduled between September 2002 and November 2004, which had three objectives: (i) to assess the performance of the upgraded activated sludge processes, and make recommendations for the appropriate upgrading of the existing activated sludge process; (ii) to explore the feasibility to employ laboratory experimentation to simulate the performance of full-scale process; and (iii) to use mathematical modeling calibrated with the measured data obtained from laboratory experimentation to study the performance and design of the full-scale system. Obviously, the second and third objectives, which relate to the scale-up and down of the activated sludge process, have a wide impact on reducing the cost and time spent on process development.

Comprehensive studies were carried out during the project period including: (i) site investigation of the activated sludge processes in the three water reclamation plants (WRPs) in Singapore; (ii) laboratory experimentation to simulate the performance of full-scale activated sludge process and to optimize the process; and (iii) mathematical modeling and simulation by using Activated Sludge Model No. 1 (ASM No. 1). The outcomes of the project were encouraging and gave rise to the contents of the book.

The study was performed by a joint team comprising Singapore Utilities International (SUI) and the Public Utilities Board (PUB), Singapore. The SUI team consists: Cao Ye Shi, (Project Manager and Principal Investigator), Ang Chee Meng, Kandiah S. Raajeevan and Zhao Wei. Cao was responsible for the entire project design, development of the methodologies and approaches and sampling programme, design of the laboratory-scale process and experiment, data analysis, road map of the mathematical modeling and simulation, and final report writing. Ang and Raajeevan carried out the site and laboratory work, data interpretation and modeling and simulation. Zhao participated in the initial site and laboratory work. The team members of the PUB side include: Wah Yuen Long (Project Manager), Benny Seah, Kan Lock Meng (PUB Headquarters), Teo Kai Hong, Kwok Wing Onn and Edgar Wong (Bedok WRP), Lee Kah Seng, Noraini Z. A., Kwok Khoo Sin, Lau Choon Leng (Kranji WRP), and Kwok Bee Hong, William Phay, Lim Hong Tow, Mokhtar bin Hashim, Lim Sui Sui, Chua Shi Yong (Seletar WRP), etc. Wah participated in and advised on the project planning and implementation, and led the PUB team in information collection, site investigation and report preparation. His vast experience on and deep understanding of sewage treatment contributed enormously to keep the study application oriented. Seah and Kan spent much effort in coordinating the PUB and SUI teams during the project.

The whole book consists of six chapters. Chapter 1 highlights the background of the project and the methodologies and approaches adopted in the study, which is helpful for the reader to understand the concepts and approaches of the investigation and the organization of the book. Chapters 2 and 3 are on full-scale process investigation. Chapter 4 describes the laboratory experimentation, where the principles of scale-down adopted in activated sludge process were defined and applied. Chapter 5 is on modeling and simulation, which involves almost all the results and knowledge presented in the preceding chapters. Chapter 6 is a summary of the findings.

The manuscript was prepared by me along with active participation from the three co-authors in discussions and preparation of the manuscript. Wah advised on the framework and structure of the manuscript and provided insightful remarks; Ang passed through and contributed to the editing and formatting of the manuscript; and Raajaveen commented on Chapter 4.

During the preparation of this manuscript, pictures from several periods of my life often resurfaced: in the 1970s, when I was studying chemical engineering during my university days in Shanghai, 'scale-up' came up repeatedly in my mind; from the 1980s to 1990s, when I studied environmental science and technology in Delft, a picturesque student town, where I learned the approaches of 'scale-down' in chemical and fermentation processes; and since 2002, my career with SUI/PUB in Singapore, whereby the knowledge was applied throughout the project mentioned above to explore the challenge "*from Flask to Field in wastewater treatment*". Time flies! My sincere gratitude goes to colleagues, friends and my family. In addition, special mentions are reserved for the PUB and SUI for their trust in providing a platform for this piece of work and their kindness in permitting the release of the materials for this publication. Finally, I wish that readers find the contents interesting and useful.

CAO Ye Shi
Singapore
January 2008

About the authors

CAO YE SHI

Dr Cao is a Senior Research Scientist at the Centre for Advanced Water Technology, a division of Singapore Utilities International (a wholly owned subsidiary of Public Utilities Board (PUB), Singapore). He is a water pollution control and water quality specialist with over thirty years work experience. He is particularly interested in biological wastewater treatment focusing on process development and optimization. Currently, he is working on nutrient removal, integrated anaerobic and aerobic sewage treatment, water reuse using activated sludge and membrane coupled process. Both his University Diploma and Master Degree of Engineering are in Chemical Engineering. He studied Environmental Science and Technology in Delft, and received his Ph.D. degree from Delft University of Technology, the Netherlands. He is also a consultant with the European Commission and The World Bank.

WAH YUEN LONG

Mr Wah Yuen Long is the Director of Water Reclamation (Plants) Department of PUB, the water agency of Singapore. He joined the then Sewerage Department of the Ministry of the Environment in June 1977 as an Engineer after graduating from the University of Singapore with a degree in Civil Engineering. Over the next few years Mr Wah worked in various municipal wastewater treatment plants. In 1979 he commissioned and then managed the Bedok Water Reclamation Plant. Mr Wah then went for studies and obtained his Master in Industrial and Systems Engineering from the University of Southern California in 1984. Subsequently, he continued to be involved in the operation and maintenance of wastewater treatment plants followed by postings to handle the planning and design and construction supervision of wastewater projects. Notable projects are the covering and odour control of 4 wastewater treatment plants and their expansion using compact designs. He was involved in the conceptualization, planning, detailed design and construction of the S$3.65 billion Deep Tunnel Sewerage System particularly the 800,000 cubic metres per day Changi Water Reclamation Plant and Outfall. Since July 2004, he has been the Director of Water Reclamation (Plants) Department managing all the municipal wastewater treatment plants in Singapore.

ANG CHEE MENG

Mr Ang Chee Meng is an Assistant Scientist at the Centre for Advanced Water Technology (CAWT), a division of Singapore Utilities International (SUI), which is a wholly owned subsidiary of Public Utilities Board (PUB), Singapore. He was awarded the Degree of Bachelor of Engineering (Environmental) from the National University of Singapore (NUS) in 2002 and joined SUI after his graduation. He has been a key member of a series of applied research projects. His project references include the Public Utilities Board (PUB) of Singapore, FujiHunt Photographic Chemicals Pte Ltd and Baxter Singapore Pte Ltd, etc. He is currently involved in several PUB-SUI joint projects on nutrient removal and water reuse by integrated activated sludge and membrane bioreactor system, etc.

KANDIAH SANMUGARATNAM RAAJEEVAN

Mr Raajeevan completed his Bachelor Degree in Civil Engineering at the University of Peradeniya (Sri Lanka) and received his Master Degree in Engineering by Research (Environmental) at the National University of Singapore. He began his professional career as an Assistant Scientist at Singapore Utilities International Pte Ltd (a wholly owned subsidiary of PUB) where he was involved in the process design & research in upgrading the activated sludge process of the Singapore water reclamation plants for NEWater production and several other water and wastewater projects. Currently, he is a Lead Engineer in the Commercial & Proposal Department of Keppel Seghers Engineering Singapore Pte Ltd and leads the Process Design and Engineering for water and wastewater Tenders.

Nomenclature

ABBREVIATIONS

AER	Aerobic
ALK	Alkalinity
ANA	Anaerobic
ANO	Anoxic
APHA	American Public Health Association
AS	Activated Sludge
ASM	Activated Sludge Model
ASP	Activated Sludge Process
ASPIRE	Asia Pacific Region
AU	Aeration Unit
AUR	Ammonia Uptake Rate
AVF	Anoxic Volume Fraction
BCOD	Biodegradable Chemical Oxygen Demand
BNR	Biological Nitrogen Removal
CAWT	Centre for Advanced Water Technology
COD	Chemical Oxygen Demand

CSTR   Continuous Stirred Tank Reactor

DO   Dissolved Oxygen

DP   Denitrification Potential of Wastewater

EC   European Commission

EFF   Effluent

FC   Final Clarifier

F/M   Food to Microorganism Ratio

GAO   Glycogen Accumulating Organism

GPS   General Purpose Simulation

HRT   Hydraulic Retention Time

IAWPRC   International Association on Water Pollution Research and Control

IWA   International Water Association

LE   Ludzack-Ettinger

MEWR   Ministry of Environment and Water Resource

MLE   Modified Ludzack-Ettinger

MLR   Mixed Liquor Recirculation

MLSS   Mixed Liquor Suspended Solids

MLVSS   Mixed Liquor Volatile Suspended Solids

N   Nitrogen

NUR   Nitrate Uptake Rate

OD   Oxygen Demand

OUD   Oxygen Uptake Demand

OUR   Oxygen Uptake Rate

P   Phosphorus

PAO   Phosphorus Accumulating Organism

PDP   Plant Denitrification Potential

PF   Peak Factor

PFR   Plug Flow Reactor

PHB   Poly-Beta Hydroxylbutyrate

PUB   Public Utilities Board

RAS   Return Activated Sludge

RBCOD   Readily Biodegradable Chemical Oxygen Demand

RHBCOD   Rapidly Hydrolysable Biodegradable Chemical Oxygen Demand

RIET   Regional Institute of Environment Technology

RS	Raw Sewage
SCOD	Soluble Chemical Oxygen Demand
SFAS	Step Feed Activated Sludge
SHBCOD	Slowly Hydrolysable Biodegradable Chemical Oxygen Demand
SND	Simultaneous Nitrification Denitrification
SNR	Specific Nitrification Removal Rate
SDR	Specific Denitrification Removal Rate
SP	Sampling Point
SR	Step Ratio
SRT	Solids Retention Time
SS	Settled Sewage
STP	Standard Temperature Pressure
SUI	Singapore Utilities International Pte Ltd
SVI	Sludge Volume Index
TKN	Total Kjeldahl Nitrogen
TN	Total Nitrogen
TOC	Total Organic Carbon
TP	Total Phosphorus
TSS	Total Suspended Solids
USEPA	United States Environmental Protection Agency
VFA	Volatile Fatty Acid
VOC	Volatile Organic Compound
VSS	Volatile Suspended Solids
WAS	Waste Activated Sludge
WEF	Water Environment Federation
WERF	Water Environment Research Foundation
WHO	World Health Organization
WRP	Water Reclamation Plant
WWTP	Wastewater Treatment Plant
XCOD	Particulate Chemical Oxygen Demand

SYMBOLS

b_A	Decay coefficient for autotrophs	d^{-1}
b_H	Decay coefficient for heterotrophs	d^{-1}
BOD_5	5-day biochemical oxygen demand	$mg\ BOD_5\ l^{-1}$
BOD_U	Ultimate biochemical oxygen demand	$mg\ BOD_U\ l^{-1}$
B_X	Carbonaceous sludge loading rate (F/M ratio)	$g\ COD(g\ VSS)^{-1}\ d^{-1}$
C_i	Concentration of ith component	$mg\ l^{-1}$
C_{IN}	Concentration at inlet	$mg\ l^{-1}$
$C_{f/r,t}$	Influent-based concentration removal or formation during a time interval	$mg\ l^{-1}$
C_{OUT}	Concentration at outlet	$mg\ l^{-1}$
$CBOD_5$	Carbonaceous biochemical oxygen demand	$kg\ COD\ d^{-1}$
COD_{IN}	Influent COD	$kg\ COD\ d^{-1}$
COD_{DN}	Denitrified COD	$kg\ COD\ d^{-1}$
COD_{OUT}	Effluent COD	$kg\ COD\ d^{-1}$
COD_{WAS}	Waste sludge COD	$kg\ COD\ d^{-1}$
f_B	Biodegradable fraction of sewage COD	-
f'_D	Fraction of biomass leading to particulate products	-
fnh	Fraction of ionized NH_4^+-N in TKN	-
$frsi$	Fraction of inert soluble organics in soluble COD	-
$frxh$	Fraction of heterotrophic biomass in particulate COD	-
$frxs$	Fraction of slowly biodegradable particulate COD in particulate COD	-
fxn	Fraction of particulate organic nitrogen in total organic nitrogen	-
icv	Particulate COD to VSS ratio	$mg\ COD\ (g\ VSS)^{-1}$
$i_{N/XB}$	Mass of N per mass of COD in biomass	$g\ N\ (g\ cell\ COD)^{-1}$
$i_{N/XD}$	Mass of N per mass of COD in particulate products from biomass decay	$g\ N\ (g\ cell\ COD)^{-1}$
ivt	VSS to TSS ratio	$mg\ VSS\ (g\ TSS)^{-1}$
K_{10}	Kinetic constant at $10\,^{\circ}C$	-
K_{20}	Kinetic constant at $20\,^{\circ}C$	-
k_a	Ammonification rate	$l\ (mg\ COD)^{-1}\ d^{-1}$

k_h	Maximum specific hydrolysis rate of SBCOD	g COD (g cell COD)$^{-1}$ d^{-1}
K_{NH}	NH$_4^+$-N half saturation coefficient for autotrophic biomass	mg COD l^{-1}
K_{NO}	NO$_3^-$-N half saturation coefficient for denitrifying heterotrophic biomass	mg N l^{-1}
$K_{O,A}$	Oxygen half saturation coefficient for autotrophic biomass	mg COD l^{-1}
$K_{O,H}$	Oxygen half saturation coefficient for heterotrophic biomass	mg COD l^{-1}
K_S	RBCOD substrate half saturation coefficient for heterotrophic biomass	mg COD l^{-1}
K_T	Kinetic constant at T $^\circ$C	-
K_X	Hydrolysis saturation constant	g COD (g cell COD)$^{-1}$
$MLVSS_{ANO}$	Volatile mixed liquor suspension solid concentration in anoxic compartment	mg COD l^{-1}
N_{ASSIM}	Nitrogen assimilated	mg N l^{-1}
N_{DN}	NO$_3^-$-N denitrified	mg N l^{-1}
NH$_4^+$-N$_{IN}$	Influent NH$_4^+$-N concentration	mg N l^{-1}
NH$_4^+$-N$_{OUT}$	Effluent NH$_4^+$-N concentration	mg N l^{-1}
NH$_4^+$-N$_{EFF}$	NH$_4^+$-N concentration in final effluent	mg N l^{-1}
Q_{AVG}	Average influent sewage flow rate	l s^{-1}
NO$_3^-$-N$_{IN}$	Influent NO$_3^-$-N concentration	mg N l^{-1}
NO$_3^-$-N$_{OUT}$	Effluent NO$_3^-$-N concentration	mg N l^{-1}
Q_I	Influent sewage flow during the time internal	m^3 d^{-1}
Q_{IN}	Influent sewage flow rate	l d^{-1}
Q_{OUT}	Outlet sewage flow rate	l d^{-1}
RO_H	Heterotrophic oxygen requirement (consumption)	kg O$_2$ d^{-1}
RO_A	Autotrophic oxygen requirement (consumption)	kg O$_2$ d^{-1}
r_D	Specific denitrification rate	mg N (g VSS)$^{-1}$ h^{-1}
r_{DSS}	Specific denitrification rate on S_S	mg N (g VSS)$^{-1}$ h^{-1}
r_{DXS2}	Specific denitrification rate on X_{S1}	mg N (g VSS)$^{-1}$ h^{-1}
r_{DXS2}	Specific denitrification rate on X_{S2}	mg N (g VSS)$^{-1}$ h^{-1}
r_i	Process rate of ith component in ASM No. 1 matrix	g COD l^{-1} d^{-1}
r_j	jth process rate in ASM No. 1 matrix	g COD l^{-1} d^{-1}
S	Substrate concentration	mg COD l^{-1}

S_{ALK}	Alkalinity concentration	mg as $CaCO_3$ l^{-1}
$SCOD_{INF}$	Influent soluble COD	mg l^{-1}
$SCOD_{EFF}$	Soluble COD in the effluent	mg l^{-1}
S_I	Soluble inert organic concentration	mg COD l^{-1}
S_i	COD substrate concentration	mg COD l^{-1}
S_{NH}	NH_4^+-N concentration	mg N l^{-1}
S_{NO}	NO_3^--N concentration	mg N l^{-1}
SNR_{AVG}	Average specific nitrification removal rate	g NH_4^+-N (kg VSS)$^{-1}$ h^{-1}
S_{NS}	Soluble biodegradable organic-nitrogen concentration	mg N l^{-1}
S_O	Dissolved oxygen concentration	mg O_2 l^{-1}
SRT_{AER}	Aerobic solids retention time	d
SRT_{ANO}	Anoxic solids retention time	d
SRT_{TOT}	Total solids retention time	d
S_S	Readily biodegradable COD concentration	mg COD l^{-1}
TKN_{IN}	Influent total kjeldahl nitrogen concentration	mg N l^{-1}
TKN_{OUT}	Effluent total kjeldahl nitrogen concentration	mg N l^{-1}
TN_{IN}	Influent total nitrogen concentration	mg N l^{-1}
V_{AER}	Aerobic reactor volume	l
V_{ANO}	Anoxic reactor volume	l
V_{TOT}	Total reactor volume	l
X_B	Concentration of biomass	g VSS l^{-1}
$X_{B,A}$	Autotrophic biomass concentration	mg COD l^{-1}
$X_{B,H}$	Heterotrophic biomass concentration	mg COD l^{-1}
X_{ENDO}	COD inside cells for respiration	mg COD l^{-1}
X_I	Particulate inert organic concentration	mg COD l^{-1}
X_{NS}	Particulate biodegradable organic-nitrogen concentration	mg N l^{-1}
X_S	Slowly biodegradable COD	mg COD l^{-1}
X_{S1}	Rapidly hydrolysable biodegradable COD	mg COD l^{-1}
X_{S2}	Slowly hydrolysable biodegradable COD	mg COD l^{-1}
X_U	Unbiodegradable particulates from cell decay	mg COD l^{-1}
Y_A	Autotrophic biomass yield	g cell COD (mg COD)$^{-1}$

Y_H	Heterotrophic biomass yield	g cell COD (mg COD)$^{-1}$
Y_{HD}	Heterotrophic denitrifier yield	g cell COD (mg COD)$^{-1}$
T	Temperature	$^{\circ}C$
ΔC	Concentration change	mg l^{-1}
$\Delta C/\Delta t$	Rate of concentration change with time	mg l^{-1}s^{-1}
ΔM	Mass change	mg
$[\Delta M/\Delta t]_{f / r,t}$	Rate of mass formation or removal during a time interval	mg s^{-1}
Δt	Time interval	h
ΣQ_i	Total daily influent sewage flow	m^3 d^{-1}
$\hat{\mu}_A$	Maximum autotrophic specific growth rate	d^{-1}
$\hat{\mu}_H$	Maximum heterotrophic denitrifier specific growth rate	d^{-1}
θ	Temperature correction coefficient	-
Θ_{XA}	Necessary aerobic sludge age for nitrification	d
Ψ_{ij}	Coefficient of ith component and jth process in ASM No. 1 matrix	-
η_h	Correction factor for heterotrophic bacteria growth and denitrification associated with anoxic condition	-
η_y	Correction factor for hydrolysis of slowly biodegrdable organic matter under anoxic condition	-

1

General introduction

1.1 BACKGROUND OF THE STUDY

Singapore is a city state with a land area of 704 km^2 and a population of more than 4.4 million. It is highly industrialized and has well established business and financial services. As a highly urbanized country, but lacking natural resources, Singapore is facing a serious shortage of water resources. Over the years, it has strived to explore new water resources, and remarkable progresses have been achieved (Leitmann, 2000; Tan, 2004). The NEWater programme, which was initiated in a demonstration plant in 2000, produces potable grade water from treated secondary effluent by further purification with the dual-membrane technology. The quality of NEWater meets the drinking water standards of the World Health Organization (WHO), European Commission (EC) and United States Environment Protection Agency (USEPA). NEWater is used for non-direct potable reuse mainly in wafer fabrication plants and cooling towers etc., and discharged into reservoirs to increase the water supply capacity. Presently, the total production capacity of NEWater in Singapore is 260 000 m^3 d^{-1}, which is derived from the four NEWater plants in four water reclamation plants (WRPs) (sewage treatment plants) i.e., Bedok, Kranji, Seletar and Ulu Pandan WRPs. NEWater constitutes one of the national water taps for Singapore (Lim, 2005; MEWR, 2005; MEWR, 2006) and marks a milestone achievement in the field of water reuse in the world (USEPA, 2004).

Due to the stringent water quality requirements of NEWater production, the secondary effluent quality from the WRPs must be further improved to meet the feed water quality requirements of the NEWater plants. Among the specific parameters, the ammonia-nitrogen concentration should be < 5 mg NH_4^+-N l^{-1}. The original conventional activated sludge process in Singapore does not facilitate ammonia removal and, thus, the ammonia concentration of the final effluent is usually in the range between 20 and 25 mg l^{-1} or higher, except, in one WRP where a Ludzack-Ettinger (LE) process (Grady *et al.*, 1999) was adopted to improve sludge settleability (Cao *et al.*, 2005a).

To cope with the feed water quality requirements of the NEWater plants, upgrading of the activated sludge processes was undertaken at the four WRPs mentioned by incorporating the biological nitrogen removal (BNR) process in the existing activated sludge processes. The process adopted in the four WRPs is, in principle, a modified Ludzack-Ettinger (MLE) process (Barnard, 1973) where a mixed liquor recirculation (MLR) stream is recycled from the final aerobic compartment of the activated sludge process to the anoxic compartment located at the head of the process.

A project titled "Application of Pre-Denitrification Technology in Bedok, Kranji, and Seletar Water Reclamation Plants" (Contract Ref No: SUI/2001/030) was awarded to the Centre of Advanced Water Technology (CAWT), a division of Singapore Utilities International Pte Ltd (SUI), in 2002. Investigation into the incorporation of BNR into the activated sludge processes in the three WRPs in Singapore was one of the main tasks of the project. At that time, the upgrading of the activated sludge processes in Bedok (Figure 1.1(a)) and Kranji (Figure 1.1(b)) WRPs was almost complete, and the design for the upgrading in Seletar WRP (Figure 1.1(c)) was nearly finalized, while upgrading in Ulu Pandan WRP was not initiated yet. Table 1.1 presents the relevant information of the three WRPs and the selected activated sludge trains including the capacity and processes. The total treatment capacity of the three WRPs is 630 000 m^3 d^{-1}, which accounts for almost half of the total municipal sewage in Singapore, while the capacity of the activated sludge process investigated in this project was 198 000 m^3 d^{-1}. The project duration was two years and was completed in 2004.

Figure 1.1(a). Birds-eye view of Bedok Water Reclamation Plant (courtesy of PUB, Singapore).

Figure 1.1(b). Birds-eye view of Kranji Water Reclamation Plant (courtesy of PUB, Singapore).

Figure 1.1(c). Birds-eye view of Seletar Water Reclamation Plant (courtesy of PUB, Singapore).

Table 1.1. Details of the three WRPs and the activated sludge trains selected for investigation.

WRP	Design capacity ($m^3 d^{-1}$)	Activated sludge process trains studied		
		Train location	Capacity ($m^3 d^{-1}$)	Process description
Bedok	232 000	Phase IV (2 trains)	58 000	AU7: MLE ASP AU8: LE ASP Anoxic volumetric ratio: 25% of total volume
Kranji	151 000	Phase III (6 trains)	75 000	MLE ASP Anoxic volumetric ratio: 20% of total volume
Seletar	247 000	Battery B (3 trains)	65 000	Conventional ASP with three lanes

1.2 OBJECTIVES OF THE STUDY

Despite BNR being widely applied in Europe and Northern America since the early 1980s, not much full-scale experience, with the exception of Japan, can be found in Asia. Little practical information is available on BNR in warm climates, in which nitrification is temperature sensitive (Stanier *et al.*, 1986; Henze *et al.*, 1997). Therefore, a number of challenges are faced in applying and incorporating BNR into the existing activated sludge processes in Singapore.

The principal objective identified was 'to ensure the secondary effluent meets the required quality specifications through the studying of the BNR activated sludge processes in the three WRPs, and to develop the design guidelines of the BNR activated sludge process in warm climates'.

The specific objectives were further refined as follow:

i   To provide information that is helpful in the operation of the BNR activated sludge process, and to evaluate the design of the upgraded activated sludge processes in the three WRPs. For this, detailed study on the performance and efficiency of the full-scale MLE/LE activated sludge processes in the three WRPs, including influent characterization and effluent quality, and the functions of the anoxic, aerobic compartments, final clarifier and RAS of the activated sludge process were carried out. A quantified description on the reaction rates and mass flow was then derived;

ii   To make recommendations on the optimal upgrading of the BNR activated sludge process in warm climates. Given the lack of experience, the existing activated sludge process upgraded might not necessarily be the most appropriate, and there is room for improvement. The outcomes of the studies could be helpful to future upgrading and accommodation of BNR into the existing activated sludge process in Singapore and the region;

iii To explore the feasibility of using a laboratory-scale system for the simulation and study of the performance of the full-scale activated sludge process. Efforts were made to explore the possibility of using a 5-6 litres laboratory-scale activated sludge system, for the simulation and prediction of the performance of the full-scale activated sludge process with a volume ranging between 6 000 and 8 700 m^3: a challenge from flask to field. To meet this purpose, the principle of scale–down, which applies to the activated sludge process, was developed and adopted in this study; and iv To develop a capacity that enables simple, reliable and modeling-aided design of BNR activated sludge processes in warm climate. Three sub-tasks were defined and undertaken: (a) development of an influent model that allows the use of conventional monitoring parameters for modeling purpose; (b) performing parameter identification based on the laboratory experimental results; and (c) keeping the number of parameters needed for identification to a minimum while adopting the default values to the maximum extent.

The components of (i) and (ii), which were carried out by holistic and systematic scale-down and scale-up approaches, are related to the geographical location. Obviously, the components of (iii) and (iv) have wide implication in process development and modeling, and are not limited to warm climates only.

1.3 METHODOLOGIES AND APPROACHES

Given the tasks and features of the project, this study is applied research oriented. The outcomes are expected to answer questions arising in practice and to be applicable to the design and operation of the full-scale system in a warm climate. A thorough understanding of the system performance under the diurnal feed conditions on site is necessary and critical. Therefore, comprehensive work on detailed information collection on key design and operation parameters, and analysis of the full-scale system performance under dynamic conditions were carried out initially. On the other hand, laboratory-scale systems were established and operated to carry out studies on the effects of various influencing factors such as sewage characterization, system configuration and operation parameters etc.

Holistic and systematic approaches shown in Figure 1.2, which have been applied successfully in chemical and food technologies (Pase, 1980; Kossen et al., 1985; and Cao, 1986) but little in wastewater treatment process, were adopted in the study.

'Scale-down' is the procedure of designing and performing the laboratory experiment based on the regime analysis of the full-scale (production-scale) process; while 'scale-up' is the procedure of applying the results of scale-down experimentation and modeling to the design and optimization of the full-scale (production-scale) process.

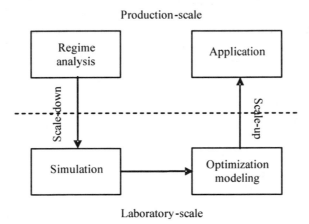

Figure 1.2. Scale-down and scale-up procedures (adapted from Sweere et al., 1987. Copyright © Elsevier Science with permission).

The project activities began with the investigation of the activated sludge processes in the three WRPs (Table 1.1). The main task was to collect this information and parameters, which were related to the 'ruling regime' governing the performance of an activated sludge process such as sewage and sludge characterization, diurnal hydraulic flow, solid retention time (SRT), hydraulic retention time (HRT), process configuration, recycle ratio, operational temperature, etc. Study of the performance of the full-scale activated sludge process under the site conditions was conducted.

The laboratory-scale activated sludge process and system, which shared the same 'ruling regime' as the full-scale process, was designed and established to simulate the performance of the full-scale system and process, and investigate the effects of various influencing factors and conditions for optimization in operation and design. The Activated Sludge Model No. 1 (ASM No. 1) (Henze *et al.*, 1987) was selected for modeling and simulation. The model parameters were calibrated with the measured data obtained from the laboratory-scale activated sludge system, and verified by the measured data obtained from both laboratory- and full-scale systems. Both the laboratory experiment and modeling were used to study the performance and optimization of the full-scale BNR activated sludge process. Scale-up of the BNR activated sludge process in a warm climate was carried out by both laboratory simulation and mathematical modeling.

In essence, three blocks constitute the main works of the study. Each of them, their components and relationships, are introduced as follow and presented in Figure 1.3.

i   **Full-scale (production) investigation:** information collection of the process and system design including capacity, SRT, HRT, configuration, sizes, aeration, and RAS and MLR ratios etc. Detailed investigation on the performance and efficiency of the whole activated sludge process as well as the individual compartment under the site conditions, was carried out based on characterization of the sewage and sludge and the outcomes of a 24-h sampling programme;

ii **Laboratory experimentation:** development of the scale-down principle and definition of the parameters of 'regime analysis' in the activated sludge process, design and establishment of laboratory-scale system, simulation of the performance of the site process and system, and experimentation with new process configuration and operation mode for optimization purpose; and

iii **Mathematical modeling:** development of a simple influent input model which allows the use of regular monitoring data in modeling based on feed sewage characterization; parameter identification by using the data obtained from laboratory experimentation and full-scale investigation; simulation of the performance under various conditions on site, and development of guidelines for upgrading and design of the BNR activated sludge process in warm climate.

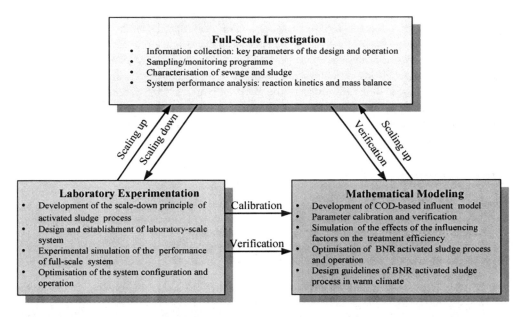

Figure 1.3. Three blocks, their components and relationships.

1.4 STRUCTURE OF THE BOOK

This book consists of six chapters. The main contents of each chapter are outlined below.

Chapter 1 highlights the background, objectives, methodologies and approaches adopted in the study, and structure of the book.

Chapter 2 introduces the detailed characterization of sewage and sludge including diurnal flow and mass loads, conventional parameters and COD fractions, and spatial nitrification and denitrification activities. This information was used in the design of the influent flow in the laboratory experiment and development of the influent model adopted for modeling.

Chapter 3 presents the results of the detailed investigation of the full-scale activated sludge process under dynamic state in one water reclamation plant including the results of a 24-h site sampling programme for which grab samples of the influent, effluent, RAS and individual anoxic and aerobic compartments were collected. Based on these data, COD removal, nitrification and denitrification in the anoxic and aerobic compartments, kinetics, mass balance and aeration energy consumption were calculated and studied. These results were used in the design of the laboratory-scale reactor system and the parameter verification of ASM No. 1.

Chapter 4 describes the parameters under three categories with respect to 'regime analysis' in the scale-down of activated sludge process, the design of laboratory experiment, the results of the simulation of the full-scale activated sludge performance, and optimization of the BNR activated sludge process in warm climate. These results contributed to the optimization and design of BNR activated sludge process in warm climate and parameter calibration and verification for ASM No. 1.

Chapter 5 introduces modeling of BNR activated sludge process under Singapore conditions using ASM No. 1. A COD-based influent model was developed based on the results of the sewage characterization. Data obtained from the laboratory-scale investigation were adopted in parameter estimation while both data from the laboratory- and full-scale investigation were adopted for verification. The calibrated model was applied to investigation of issues that are not easily studied in laboratory or on site, the factors relevant to BNR design, and appropriate upgrading of the existing BNR activated sludge process under Singapore conditions. Recommendations on upgrading the existing activated sludge process to incorporate BNR and design of the BNR activated sludge process in warm climate were presented accordingly.

Chapter 6 summarizes the major findings and conclusions of the individual chapters and of the study as a whole.

REFERENCES

Barnard J. (1973) Biological Denitrification. Wat. Pollut. Cont. **72**, 705-717.

Cao Yeshi (1986) Scale-Up and Down of Bioreactor. Final Report, International Institute of Hydraulic and Environmental Engineering (IHE). Delft, the Netherlands.

Cao Yeshi, Teo Kai Hong, Yuen Weng Ah, Wah Yuen Long and Benny Seah (2005a) Performance Analysis of Anoxic Selector in Upgrading Activated Sludge Process in Tropical Climate. Wat. Sci. Tech. 52(4/12), 27-37.

Gragy C.P.L., Daigger G. T. and Lim H.C. (1999) Biological Wastewater Treatment, 2nd ed. Marcel Dekker. New York.

Henze M., Grady C. P.L., Gujer W., Marais, G. v. R. and Matsuo, T. (1987). Activated Sludge Model No. 1, IAWPRC Sci. and Tech. Report No. 1, IAWPRC, London.

Henze M., Harremoës P., Janseen J. and Arvin E. (1997) Wastewater Treatment: Biological and Chemical Process, 2nd ed, Springer, Berlin.

Kossen N.W.F and Oosterhuis N.M.G. (1985) Modeling and Scaling-Up of Biorectors. In Biotechnology, (eds. Rehm H.J. and Reed G.), Vol. 2, 572-605, VCH Verlaggesellschaft, Weinheim.

Leitmann J. (2000) Integrating the Environment in Urban Development: Singapore as A Model of Good Practice, Working Paper Series No. 7. The World Bank. Washington D.C., USA.

Lim Chiow Giap (2005) NEWater: Closing the Water Loop. 1st IWA-ASPIRE (Asia-Pacific Regional Group) Conference & Exhibition, 10th - 15th July, 2005, Singapore.

Pase G. W. (1980) Adv. School on Microbial and Biotechnology. Queen Elizabeth Coll. London. 14-20 September 1980.

MEWR (2005) Water Strategy in Singapore, (http://www.pub.gov.sg).

MEWR (2006) The Singapore Green Plan 2012, 2006 ed.

Stanier R., Ingraham J., Wheelis M. and Painter P. (1986) General Microbiology, 5th ed, Prentice-Hall, Englewood Cliffs, New Jersey, USA.

Sweere A.P.J., Luyben K. Ch. A. M. and Kossen W.F. (1987) Regime Analysis and Scale-Down: Tolls to Investigate the Performance of Bioreactors. Enzyme Microb Tech., Vol. 9.

Tan Gee Paw (2004) After the NEWater, There are New Challenges, in: PUB Annual Report 2004.

USEPA (2004) Guidelines for Water Reuse, EPA/625/R-04/108, Washington D.C.

2

Characterization of settled sewage and activated sludge

2.1 INTRODUCTION

Wastewater characteristic is one of the fundamental factors influencing process selection, design, improvement and optimization of wastewater treatment especially for nutrient removal and modeling. A series of biological, chemical and physical processes occur in sewers during transportation which affect the sewage characteristics prior to treatment plants. Studies have been conducted on the processes in sewers (Nelson *et al.*, 1992; Cao, 1994; Raunkjær *et al.*, 1995) and interactions between sewer and wastewater treatment plants (Langeveld, 2004) but little information is available on sewage transformation in sewers and sewage characteristics in warm climates. In Singapore, more than 85% of municipal sewage comes from domestic sources and is collected through a separate gravity flow sewerage system. Given the importance of sewage and sludge characteristics, detailed characterization of the sewage and activated sludge of the three WRPs was made in the study with the following objectives:

   i   To understand the major features of municipal sewage in Singapore, which provides essential information for the treatability of the sewage and preliminary design of the sewage treatment under Singapore conditions;

ii   To provide information in the preparation of the feed sewage for laboratory experiment including diurnal hydraulic flow profile and composition concentration profiles; and

iii   To obtain detailed sewage composition data that is essential in the development of the COD-based influent model and data file for mathematical modeling and simulation under Singapore conditions.

2.1.1 Literature review

The methods of wastewater characterization are restricted to the analytical methods available and the application purposes of characterization. The conventional characterization of municipal sewage began in the 19th Century with early developments in urban sanitation. The development of activated sludge models over the last two decades led to a new paradigm of advanced characterization of sewage and activated sludge.

Conventional characterization. For municipal sewage treatment, the parameters of conventional characterization include hydraulic flow, COD, BOD_5, TKN, NH_4^+-N, phosphorus, alkalinity, TSS, VSS etc., and these parameters are further distinguished according to physical state such as soluble and particulate (Henze *et al.*, 1997; Metcalf and Eddy, 2003). These parameters are also used to express sewage bio-treatability or the extent of nutrient removal through BOD_5/COD, COD/TKN and COD/TP ratios (Marais, 1994; Henze *et al.*, 1997).

Advanced characterization for modeling. For the last two decades, mechanistic modeling of the activated sludge process has developed and achieved significant progresses in the professional understanding and practical applications of activated sludge process (Dold *et al.*, 1980; Henze *et al.*, 1987). Characterization of the sewage and microbial activity in the mixed liquor based on biodegradation properties has become a specific area.

In the Activated Sludge Model No. 1 (ASM No. 1), the influent COD is divided into readily biodegradable COD (S_S), inert soluble COD (S_I), slowly biodegradable particulate COD (X_S) and inert particulate COD (X_I) (Henze *et al.*, 1987). The slowly biodegradable particulate COD (X_S) was further divided into non-biomass particulate COD, heterotrophic bacteria COD ($X_{B,H}$) and autotrophic bacteria COD ($X_{B,A}$). For biological phosphorus removal, volatile fatty acids (VFAs) in sewage, phosphorus accumulating organisms (PAO), polyphosphate and cell internal storage products of PAOs etc., were introduced in the Activated Sludge Model No. 2 (ASM No. 2) (Henze *et al.*, 1995) and Activated Sludge Model No. 3 (ASM No. 3) (Gujer *et al.*, 1999).

The Nitrate Uptake Rate (NUR) and Oxygen Uptake Rate (OUR) tests are mainly used to determine the COD fractions and to study the kinetics of heterotrophs (Ekama *et al.* 1986; Çokgör, *et al.*, 1998; Vanrolleghem *et al.*, 1999) for modeling purposes. Ammonia Uptake Rate (AUR) is adopted in the determination of the kinetics of autotrophs (Kristensen *et al.*, 1992).

Application of wastewater treatment modeling including characterization of municipal sewage and sludge has become popular, mainly, in Europe and North America (Henze *et al.*, 1987; Henze, 1992; Kappeler and Gujer, 1992; Siegrist and Tschui, 1992; Daigger *et al.*, 1995; Naidoo *et al.*, 1998; Spérandio *et al.*, 2001). However, there are several areas where the picture is not totally clear. For example (a) the relationship between COD fraction and diurnal flow variation has yet to been fully studied; (b) the conventional approach i.e., to adopt the same set of kinetics for the description of nitrification and denitrification in the aerobic and anoxic zones, assuming no effects of the environment on the activities of nitrifiers and denitrifiers, has yet to be justified, although studies illustrated that an alternating environment could effect the enzyme induction (Baumann *et al.*, 1997); and, finally, (c) little results on COD fractions, kinetic parameters and stoichiometric coefficients in warm regions are reported although temperature could affect the magnitude of these components significantly (Kappeler and Gujer, 1992).

2.1.2 Focuses of the characterization

The contents presented in this chapter include: diurnal hydraulic flow, organic and nutrient mass loads, concentration profiles, COD fractions of settled sewages, specific denitrification rates corresponding to COD fractions and specific denitrification and nitrification activities of microorganisms in the activated sludge mixed liquor (Cao *et al.*, 2003; Raajeevan, 2003; Cao *et al.*, 2005b). The main focuses are as follow:

  i   diurnal hydraulic flow, carbonaceous and nutrient mass loads, concentration profiles of COD, nitrogen, phosphorus, total suspended solids and alkalinity etc. under Singapore conditions;

  ii   the relationship between the sewage COD fractions with diurnal hydraulic flow variation and comparison with the reported values; and

  iii nitrification and denitrification activities at different locations of the activated sludge process.

2.2 MATERIALS AND METHODS

2.2.1 Activated sludge processes of the three WRPs studied

Bedok WRP is located in the south-eastern part of Singapore. The design capacity is 232 000 m^3 d^{-1} but in recent years the actual daily total is 260 000 m^3 or more. There are eight in-parallel trains of the activated sludge process of similar plug-flow configuration and volume in Bedok WRP. Only one pumping station pumps the sewage into the wastewater treat plant. The hydraulic flow pattern recorded by the pumping station is, therefore, applicable to each of the activated sludge trains. Four Phases, each with two trains, are defined from the first to the eighth. The effective volume of each tank is 8 724 m^3, and for each tank the dimensions are 87 x 21.8 x 4.6 m (length x width x depth). For each train there are four compartments of equal size but only the first one has a physical partition. Phase IV, consisting of the activated sludge train 7 (AU7), which operated as an MLE process, and 8 (AU8), which operated as a LE process, were selected for study in this project. Figure 3.1 in the following chapter shows the process configuration and key design and operation parameters.

Kranji WRP is located in the northern part of Singapore. Sewage arrives at Kranji WRP by gravity via the North Woodlands and South Woodlands Trunk Sewer systems. The flow from the North Woodlands Trunk Sewer is pumped to and treated in Phase I and II, with a dry weather design flow capacity of 76 000 m^3 d^{-1}, while the flow from the South Woodlands Trunk Sewer is pumped to and treated in Phase III, with a design capacity of 75 000 m^3 d^{-1}. Thus, the total design treatment capacity of Kranji WRP is 151 000 m^3 d^{-1}. The actual flow treated was more than that. Phase III consists of six similar trains of plug-flow configuration and volume. Each tank is 56 m long, 11 m wide and 10 m deep and is partitioned into seven compartments. Presently, Phase III provides the feed to the NEWater plant, and to reduce the NH$_4^+$-N load, the concentrate stream from the dewatering process of Phase III is transferred to the activated sludge processes of Phases I and II. Phase III was selected for study in this project. Figure 5.13 shows the process configuration and key design and operation parameters.

Seletar WRP is located in the north-eastern part of Singapore. Seletar WRP was developed in two stages and has the capacity to handle an average sewage flow of 247 000 m^3 d^{-1}. The WRP has activated sludge processes developed in different phases. Phases I and II are earlier installations of the plant with a total design capacity of 114 000 m^3 d^{-1}. Phase III is a new installation with a pumping station separated from that of Phases I and II and a design capacity of 133 000 m^3 d^{-1} consisting of six similar trains. Each train has an effective volume of 9 644 m^3, and is a three-lane MLE configuration process. There are two batteries in Phase III (Batteries A and B) with three trains in each battery. Battery B was chosen for the study.

2.2.2 Sampling programme

Hydraulic flow profiles were obtained from respective site records of the pumping stations of Bedok WRP, Phase III of Kranji WRP and Seletar WRP. The flow into the particular activated sludge trains under site investigation could be calculated assuming uniform flow distribution among the activated sludge trains.

The settled sewage samples were collected at the inlet channel of the secondary treatment units. For conventional characterization, hourly or bi-hourly grab samples and samples from auto samplers were collected from the WRPs.

Conventional characterization of the settled sewages of Bedok, Kranji and Seletar WRPs was conducted. The samples were taken from the channels entering the activated sludge processes of Phase IV of Bedok WRP, Phase III of Kranji WRP and Battery B of Seletar WRP. COD fractionation by NUR respirometry was determined for the settled sewage taken from Phase IV of Bedok WRP and Phase III of Kranji WRP. Nitrification and denitrification activity determination was made only for the sludges from the anoxic, the first and third aerobic compartments and the RAS stream of AU8 of Bedok WRP (for the configuration, see Figure 3.1).

2.2.3 NUR and AUR tests

The NUR tests were made to study sewage COD fractionation and to determine the sludge denitrification potential as reported by Kujawa and Klapwijk (1999) while the AUR tests were made to determine the nitrification potential of the sludge as reported by Kristensen *et al.* (1992).

The NUR and AUR tests were conducted within two hours after sampling in order to preserve the activity of the microorganisms. The sludge samples taken from the anoxic zone and the RAS of AU8 of Bedok WRP were tested for denitrification activities prior to testing for nitrification activities. The sludge samples taken from aerobic compartments 1 and 3 were tested for nitrification activities prior to testing for denitrification activities. The environmental conditions were preserved in this way and the kinetics of BNR was therefore representative.

A predetermined sludge volume was added with acetate solution, settled sewage or NH_4^+-N solution in the NUR and AUR tests. The dissolved oxygen (DO) concentration in the NUR reactor was maintained below 0.05 mg l^{-1} while that of the AUR reactor was maintained above 2.0 mg l^{-1}. Both the NUR and AUR tests were made in 5.5 l reactors and the mixed liquor temperature and pH of each reactor were controlled at 30 ± 0.5°C and 7.2 ± 0.1, respectively. Grab samples were drawn every 6 to 10 minutes depending on the NO_3^--N or NH_4^+-N uptake rates, and were then filtered with Whattman 0.45 μm filter papers to prevent denitrification occurring in the sample bottles. The concentration profiles were recorded one hour after they were switched to anoxic and aerobic conditions, respectively, for adaptation. Samples were analyzed for NO_2^--N, NO_3^--N and NH_4^+-N. It was noted that NO_2^--N accumulation with time was very low (< 0.1 mg l^{-1}) in the NUR tests and, so, NO_2^--N was neglected in the calculation.

2.2.4 Analytical methods

The analyses of COD and SCOD were made using the Hach COD reactor following the reactor digestion method (Hach Method 8000), while the analysis of TP was made using the Hach COD reactor following the PhosVer®3 with acid persulfate digestion method (Hach Method 8190). NO_3^--N, was measured using the chromotropic acid method (Hach Method 10020), NO_2^--N measured using the diazotization method (Hach Method 8507) and NH_4^+-N measured using the nessler method (Hach Method 8038), respectively. The Hach Odyssey DR/2500 colorimeter was used for all the above methods, which are USEPA approved (Federal Register, 1980) except Hach Method 8038, which is adapted from Standard Methods (APHA, 1998). Parameter analyses of TKN, TSS, VSS and alkalinity were according to APHA (1998).

2.3 RESULTS AND DISCUSSION

2.3.1 Hydraulic flow and carbonaceous and nitrogenous mass loadings

Figures 2.1 and 2.2 show the diurnal hydraulic flow rates of Bedok WRP (8 trains) and Phase III of Kranji WRP (6 trains), respectively. The two profiles exhibited different flow patterns. For Bedok WRP, there was a peak between 12:00 and 18:00 with a peak factor of 1.5 but the presence of a second peak was not apparent. The low flow occurred between 00:00 and 05:00 with a base flow factor of 0.6. There was strong fluctuation for the flow of Phase III of Kranji WRP compared to that of Bedok WRP. Two high flows occurred, one between 10:00 and 13:00; and the other between 21:00 and 01:00. Both had peak factors of about 1.5 although values as high as 1.8 have been recorded; and sometimes the second flow peak was lower than the first. The base flow occurred between 03:00 and 07:00 with a base flow factor of about 0.2 only. The diurnal flow pattern of Battery B of Seletar WRP (3 trains) was similar to that of Phase III of Kranji WRP but the fluctuation was even stronger (Cao *et al.*, 2004a).

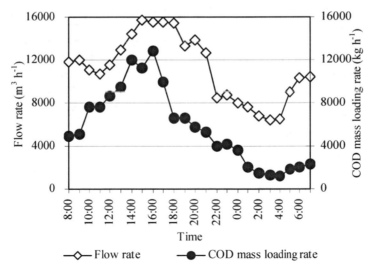

Figure 2.1. Diurnal flow and COD mass loading rates of Bedok WRP (27-28 December 2002).

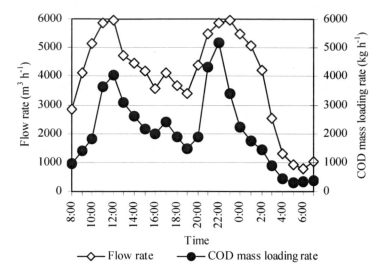

Figure 2.2. Diurnal flow and COD mass loading rates of Phase III of Kranji WRP (11-12 November 2002).

For comparison purposes, the diurnal mass loading rates of COD and NH_4^+-N were presented together with the hydraulic flow, which were calculated from the flow and COD and NH_4^+-N concentration profiles presented in Figures 2.5–2.8. The common feature (including Battery B of Seletar (Cao *et al.*, 2004a)) was that the COD mass loading rate peaks coincided with the hydraulic flow peaks although the correlation during the second flow peak was not apparent in Bedok WRP. For the ammonia nitrogen mass loading rate, the peak coincided with the first hydraulic flow peak, but there was no apparent correlation during the second flow peak. Recycling of the dewatering supernatant containing high ammonia concentration can contribute to this phenomenon. For example, in Bedok WRP, the second ammonia-nitrogen mass loading rate peak occurred between 05:00 and 07:00, which was most likely due to the recycling of the supernatant of the dewatering operation from the late night to the early morning; while in Phase III of Kranji WRP and Battery B of Seletar WRP, the dewatering supernatant was recycled to the other activated sludge process trains. The mass loading rate variation had an impact on operation. The first COD and ammonia-nitrogen mass loading peaks corresponded to a high oxygen demand, and could have caused an ammonia load shock, which would have affected the ammonia concentration in the final effluent. Both the low COD and ammonia-nitrogen mass loading rates coincided with the base flow period, which corresponded to a reduced oxygen demand as compared to the other times. The aeration capacity should be able to adjust accordingly to the extremely high and low oxygen demands during the peak and base flow periods, respectively.

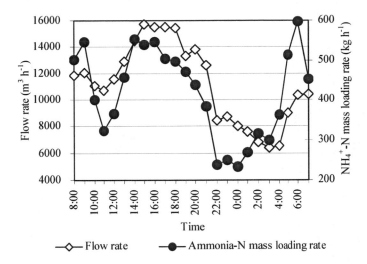

Figure 2.3. Diurnal flow and NH_4^+-N mass loading rates of Bedok WRP (27-28 December 2002).

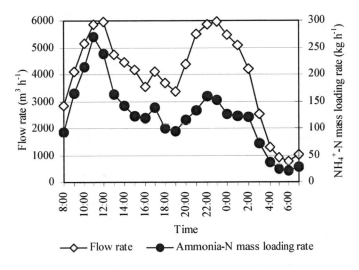

Figure 2.4. Diurnal flow and NH_4^+-N mass loading rates of Phase III of Kranji WRP (11-12 November 2002).

2.3.2 Conventional parameters and their fluctuations

The diurnal COD, SCOD, BOD_5 and TSS concentration profiles of the settled sewages of Bedok and Phase III of Kranji WRPs are shown in Figures 2.5 and 2.6, respectively. The common feature is that the concentration peaks of COD, BOD_5 and TSS coincided with the hydraulic flow peak, while the variation of SCOD with the flow was not apparent. This indicated that the increases in COD and BOD_5 during the hydraulic flow peaks were most likely due to the increase of TSS by re-suspension.

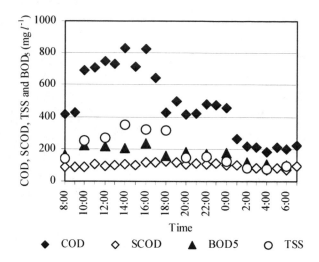

Figure 2.5. Diurnal COD, SCOD, BOD₅ and TSS concentration profiles of the settled sewage of Bedok WRP (27-28 December 2002).

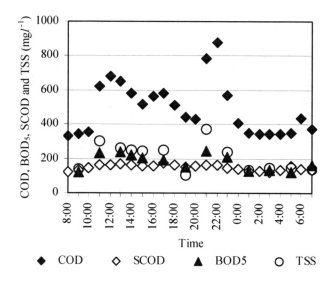

Figure 2.6. Diurnal COD, SCOD, BOD₅ and TSS concentration profiles of the settled sewage of Phase III of Kranji WRP (11-12 November 2002).

The concentration profiles of nutrient constituents including TKN, NH_4^+-N and TP of Bedok and Kranji WRPs are shown in Figures 2.7 and 2.8, respectively. For Bedok WRP, the variation of the concentration profiles of TKN and NH_4^+-N exhibited less obvious correlation with the hydraulic flow peak. However, both increased during the base flow period, which was most likely related to the recycling of the supernatant from the dewatering operation at night. Differing from Bedok WRP, the TKN and TP concentrations of Phase III of Kranji WRP increased with the daytime flow, while concentration increase was not observed during the base flow period, which was consistent with the fact that the supernatant of dewatering was not recycled into the trains during that period. In Battery B of Seletar WRP, the TKN, NH_4^+-N and TP concentrations increased during the peak flow period in the morning (data not shown) (Cao *et al.*, 2004a).

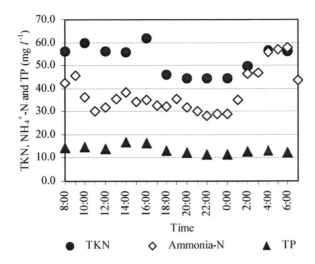

Figure 2.7. Diurnal TKN, NH_4^+-N and TP concentration profiles of the settled sewage of Bedok WRP (27-28 December 2002).

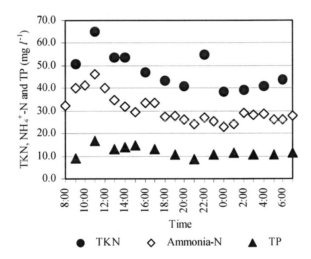

Figure 2.8. Diurnal TKN, NH_4^+-N and TP concentration profiles of the settled sewage of Phase III of Kranji WRP (11-12 November 2002).

The summary of the conventional parameter data of the settled sewages of Bedok WRP, Phase III of Kranji WRP and Battery B of Seletar WRP is compiled in Table 2.1 along with the reported data (Henze *et al.*, 1997) for comparison purpose. The respective total CODs for Bedok and Kranji WRPs were 513 and 526 mg l^{-1}, within the 'concentrated' range, while the total COD of Seletar WRP was 366 mg l^{-1}, within the 'moderate' range. However, the respective soluble CODs (SCOD) of the three WRPs were all within the 'diluted' range, while the TSS concentrations of the three WRPs were more than 200 mg l^{-1}, much higher than the 'concentrated' range. It should be noted that on some occasions, the COD of the settled sewage of Bedok and Kranji WRPs could be lower (~ 250 mg l^{-1}, Section 3.3.2) than those presented here, possibly due to differing weather conditions.

Table 2.1. Conventional parameters of settled sewage of Bedok WRP (27-28 December 2002), Kranji WRP (11-12 November, 2002), Seletar WRP (26-27 February 2004) and literature values (mg l⁻¹).

Parameter	Range			Weighted average[*]			Literature values (Henze et al., 1997)			
	Bedok	Kranji	Seletar	Bedok	Kranji	Seletar	Concentrated	Moderate	Diluted	Very diluted
COD	180 – 831	330 – 880	260 – 670	513	526	366	530	370	230	150
SCOD	68 – 123	125 – 176	86 – 157	100	151	117	300	210	130	80
BOD$_5$	97 – 236	121 – 247	55 – 78	177	193	67	250	175	110	70
TKN	44.2 – 61.9	34.6 – 65.0	34.2 – 54.0	52.6	50.0	43.0	80	50	30	20
NH$_4^+$-N	28.1 – 57.8	23.0 – 46.2	24.8 – 43.0	37.4	31.3	30.4	50	30	18	12
TN	40 – 65	32 – 60	-	53.0	51.0	-	80	50	30	20
TP	11.2 – 16.7	8.5 – 16.9	6.6 – 17.3	13.7	12.2	10.5	23	16	10	6
TSS	78 – 348	105 – 371	110 – 540	215	231	207	130	90	50	40
ALK (as CaCO$_3$)	168 – 341	147 – 227	130 – 184	217	179	148	150 – 300	150 – 300	150 – 300	150 – 300
pH	6.8 – 7.5	6.6 – 6.9	6.5 – 7.0				7 – 8	7 – 8	7 – 8	7 – 8

[*] Calculated by the equation ($\Sigma[Q_i \times C_i]/\Sigma Q_i$) where C_i and Q_i are COD and flow measured every 2 h.

The TKN concentrations of the three plants were either within or close to the 'moderate' range. The TP concentrations were between the 'diluted' and 'moderate' ranges. The weighted alkalinity concentrations of the settled sewages of the three plants varied between 148 and 217 mg (as $CaCO_3$) l^{-1}, which leant towards the lower limits of reported values (Henze et al., 1997). This was, most likely, due to the fact that all the water resources in Singapore were from surface water.

The weighted ALK/NH_4^+-N ratios were < 7.08 mg $CaCO_3$ (mg NH_4^+-N)$^{-1}$, which is the stoichiometric coefficient of alkalinity consumption per mg NH_4^+-N oxidized, The SCOD/TKN ratios were close or < 2.86 mg COD (mg NO_3^--N)$^{-1}$, which is the stoichiometric coefficient of COD consumption per mg NO_3^--N reduced, indicating that the soluble COD might be insufficient for an efficient denitrification. These values illustrated that a large anoxic reactor might be needed to achieve high denitrification efficiency under Singapore conditions.

In summary, diluted SCOD, low alkalinity and low ratios of ALK/NH_4^+-N and SCOD/TKN indicated that (a) the soluble biodegradable COD might not be sufficient for efficient nutrient removal; (b) either alkalinity addition or denitrification is necessary when nitrification is facilitated; and (c) a large reactor volume might be required for the anoxic reactor for a high denitrification efficiency. The COD/BOD_5 ratios of the settled sewages of the three WRPs calculated from the data in Table 2.1 were > 2.7, indicating a high portion ($\geq$ 20%) of inert solids in the settled sewages (Takàcs, 2006).

2.3.3 COD fractions and their fluctuations

Figures 2.9 and 2.10 show the depletion of NO_3^--N with time in NUR tests performed with the settled sewages collected from Phase IV of Bedok WRP and Phase III of Kranji WRP, respectively. In each figure, there were four distinct phases with different rates, as reported by Çokgör et al. (1998), which corresponded to the four constitutuents of COD, i.e., readily biodegradable COD (S_S), rapidly hydrolysable biodegradable COD (X_{S1}), slowly hydrolysable biodegradable COD (X_{S2}) and endogenous respiratory COD (X_{ENDO}).

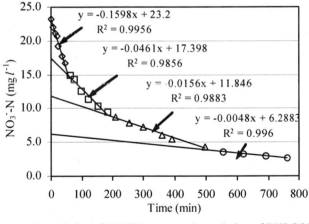

Figure 2.9. NO_3^--N reduction of the Phase IV settled sewage of the Bedok WRP during the NUR experiment (23 July 2003, peak flow, dry weather).

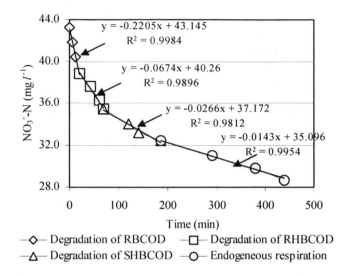

Figure 2.10. NO_3^--N reduction of the settled sewage of Phase III Kranji WRP during the NUR experiment (29 October 2002, normal flow, dry weather).

2.3.3.1 COD fractions under normal hydraulic flow

The COD fractions of the settled sewages were determined by using the methods reported by Kujawa and Klapwijk (1999). COD fractionation under normal flow conditions was carried out through the NUR test with the settled sewage collected at the time when the flow rate was close to the daily average value. For Bedok and Kranji WRPs, under the normal flow conditions, the respective total COD were 286 and 369 mg l^{-1}, and the SCOD were 131 and 127 mg l^{-1}, respectively, and are described as follows.

Readily biodegradable COD (S_S). The readily biodegradable COD (S_S) of the settled sewages of Bedok and Kranji WRPs were 51 and 50 mg l^{-1}, respectively.

Rapidly and slowly hydrolysable biodegradable COD (X_{S1} & X_{S2}). The X_{S1} of Bedok and Kranji WRPs were 21 and 49 mg l^{-1}, respectively, while the X_{S2} were 26 and 59 mg l^{-1}, respectively. The sum of X_{S1} and X_{S2} of each of the settled sewages of Bedok and Kranji WRPs, were 47 and 108 mg l^{-1}, respectively.

Slowly biodegradable COD (X_S). X_S was calculated as the sum of X_{S1}, X_{S2} and $X_{B,H}$ (heterotrophic biomass COD). Assuming $X_{B,H}$ to be 20% of the particulate COD (COD-SCOD) as adopted by GPS-X (Hydromantis Inc., 1999), the X_S of the settled sewages of Bedok and Kranji WRPs were 78 and 156 mg l^{-1}, respectively.

Inert particulate COD (X_I). X_I was calculated as the difference between COD and the sum of S_I, S_S and X_S. Taking S_I as 20 mg l^{-1} and the other COD constituents mentioned above, the X_I of the settled sewages of Bedok and Kranji WRPs under the normal flow condition were 137 and 142 mg l^{-1}, respectively.

The average values of the COD and its fractions presented in Table 2.2 are representative of the domestic settled sewage in Singapore and are compared with reported values. The nitrogen fractions were derived from the respective data of the three plants in Table 2.1.

Table 2.2. Typical characteristics of the domestic settled sewages of Europe (Henze *et al.*, 1987) and the present study.

Symbol	Unit	Denmark	Hungary	Switzerland	Singapore
COD	mg l^{-1}	515	350	220	328
S_S	mg COD l^{-1}	125	100	70	51
S_I	mg COD l^{-1}	40	30	25	20
X_S	mg COD l^{-1}	250	150	100	120
X_I	mg COD l^{-1}	100	70	25	139
S_{ND}	mg N l^{-1}	8	10	5	5
X_{ND}	mg N l^{-1}	10	15	10	16
S_{NH}	mg NH_4^+-N l^{-1}	30	30	10	32
S_{NI}	mg N l^{-1}	2	3	2	1.5
S_{NO}	mg NO_3^--N l^{-1}	0.5	1	1	0

Comparing the settled sewage of Singapore with those reported in Table 2.2, it could be found that the readily biodegradable COD (S_S) of 51 mg COD l^{-1} was less than those of the reported range between 70 and 125 mg COD l^{-1}. The slowly biodegradable COD (X_S) of 120 mg COD l^{-1} was closer to the lower boundary of the reported range between 100 and 250 mg COD l^{-1}. It was realized that, most likely, the endogenous respiratory COD (X_{ENDO}) could be part of X_{S2}, and it might have been underestimated since the duration of the NUR test of about 7.5 h was shorter than the site SRT of 10 d although the percentage of the sum of S_S and X_S over total COD was similar to the literature value (Naidoo, *et al.* 1998). Thus, X_S might have been underestimated, and ,as a consequence, the inert particulate COD (X_I) might have been overestimated although the high ratio of COD/BOD$_5$ (> 2.7) indicated a high portion of inert solid in the settled sewage as mentioned earlier (Section 2.3.2). No significant differences in nitrogen fractions were found when the Singapore data were compared with those reported.

High temperature in a warm climate, which enhances biodegradation of soluble COD and hydrolysis of particulate COD during the transportation of sewage in the sewer system, could be a major reason for the lower S_S and X_S of the settled sewage under Singapore conditions. More study may be needed to find out the causes of high X_I.

2.3.3.2 Variation of COD fractions with hydraulic flow

To study the COD fraction change with hydraulic flow conditions, NUR tests were performed with the settled sewage sampled during the peak, normal and low flow periods in the same day for both the settled sewages of Bedok and Kranji WRPs. The results are presented in Figures 2.11 and 2.12.

Readily biodegradable COD (S_S). For Bedok WRP, the readily biodegradable COD (S_S) of the settled sewage for the peak, normal and low flow periods were 66, 51 and 6 mg l^{-1}, respectively, and for Phase III of Kranji WRP, the corresponding values were 36, 50 mg l^{-1} and non-detectable during low flow. These data indicated that the S_S during the low flow period was extremely low, and the longer sewage retention time might have been the reason.

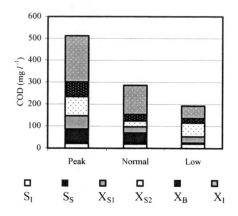 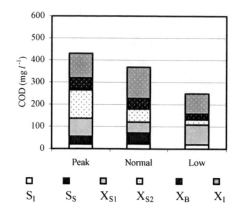

Figure 2.11. COD fractions in Bedok WRP settled sewage under peak, normal and low flow conditions.

Figure 2.12. COD fractions in Kranji WRP settled sewage under peak, normal and low flow conditions.

Rapidly hydrolysable biodegradable COD (X_{S1}). The X_{S1} of the settled sewages for the three different flow periods were 60, 27 and 28 mg l^{-1}, respectively, for Bedok WRP, and 81, 49 and 90 mg l^{-1}, correspondingly for Phase III of Kranji WRP. The higher X_{S1} values were probably related to elevated sewage COD during peak flow. Although this fraction was more susceptible to transformation in sewers compared to S_S, the reason for not observing the systematic finger print with the hydraulic flow during the base flow period might have been due to the fact that the biodegradation rate of this fraction was slower than S_S. S_S and X_{S1} are the major carbon sources for rapid denitrification and the sum of these two components during the peak, normal and low flow conditions were 126, 78 and 34 mg l^{-1} respectively for Bedok WRP, and 117, 99 and 90 mg l^{-1}, correspondingly for Phase III of Kranji WRP.

Slowly hydrolysable biodegradable COD (X_{S2}). The X_{S2} of the settled sewage of Bedok WRP for the three different flow periods were 87, 26 and 62 mg l^{-1}, respectively, and 127, 59 and 23 mg l^{-1}, correspondingly for Phase III of Kranji WRP. The X_{S2} of the peak flow period was higher than those of the normal and low flow period due to the re-suspension of particulate matter during peak flow.

The COD fractions during the three periods are summarized in Table 2.3 together with the literature values for comparison purpose. The S_S of the settled sewages of the two WRPs varied in the range between non-detectable and 66 mg l^{-1} and it was regarded as 'very diluted' when compared to the literature data. Taking X_S as the sum of X_{S1}, X_{S2} and $X_{B,H}$ (Table 2.3), the X_S of the two plants varied between 84 and 264 mg l^{-1}, which ranged between 'diluted' and 'concentrated'.

The study of COD fractions with hydraulic flow conditions revealed that COD fingerprints changed with hydraulic flow and therefore, diurnal COD fractions or concentrations should be used rather than fixed COD fractions in modeling and simulations. These data were applied in the formulation of a COD-based influent input model used in mathematical modeling (Section 5.2). Additionally, these data could be adopted in dynamic process control, e.g. on manipulating aeration supply and MLR.

Table 2.3. COD fractions under different hydraulic flow conditions of Phase IV of Bedok (B) WRP and Phase III of Kranji (K) WRP and literature values (mg l^{-1})

Hydraulic flow condition	COD		SCOD		S_I		S_S		X_{S1}		X_{S2}		$X_{B,H}$		X_I		S_S+X_{S1}		BCOD ($S_S+X_{S1}+X_{S2}+X_{B,H}$)		Lit. values (Henze et al., 1997) Range	S_S	X_S
	B*	K**	B	K	B	K	B	K	B	K	B	K	B	K	B	K	B	K	B	K			
Peak (dry)	512	430	173	149	21	20	56	36	60	81	87	127	68	56	210	110	126	117	267.0	300	Concentrated	180	290
Normal	286	369	131	127	19	22	51	50	27	49	26	59	31	48	132	141	78	99	135.0	206	Moderate	130	210
Low (dry)	193	249	87	119	20	19	6	ND	28	90	62	23	21	26	56	91	34	90	116.6	139	Diluted	75	125
																					Very diluted	50	85

* Phase IV of Bedok WRP
** Phase III of Kranji WRP
ND: non-detectable

2.3.4 Denitrification activities corresponding to COD fractions

The specific nitrate-nitrogen removal rates (r_D) corresponding to the different COD fractions were determined based on the MLVSS, nitrate-nitrogen reduction and duration of each phases in the NUR test. The specific activities (rates) of the activated sludge of Bedok and Kranji WRPs for different flow conditions are summarized in Table 2.4. The correlation between the magnitudes of the specific activities and the corresponding four COD fractions of the sludges from the two WRPs were logical and comparable. The average specific rate corresponding to S_S, r_{DSS}, was 7.20 mg N (g VSS)$^{-1}$ h^{-1}, which was higher than the reported range between 3 and 5 mg N (g VSS)$^{-1}$ h^{-1} (Naidoo et $al.$, 1998). The specific rate corresponding to X_{S1}, r_{DX1}, was 1.80 mg N (g VSS)$^{-1}$ h^{-1}, which was within the reported range between 1 and 3 mg N (g VSS)$^{-1}$ h^{-1} (Naidoo et $al.$, 1998). The specific rates corresponding to X_{S2}, r_{DX2}, and X_B, r_{Dend}, were 0.63 and 0.45 mg N (g VSS)$^{-1}$ h^{-1}, respectively, which were lower than the reported range between 1 and 2 mg N (g VSS)$^{-1}$ h^{-1} in Europe (Naidoo et $al.$, 1998). Temperature and sewage compositions were two factors resulting in the specific rate differences. The specific denitrification rate, r_{DSS}, was about four times that of r_{DX1} and more than ten times higher than r_{DX2} and r_{Dend}. These different reaction rates also illustrated that to denitrify a certain amount of NO_3^--N, a smaller anoxic zone volume would be required when S_S and X_{S1} in the settled sewage were sufficient, while a larger anoxic zone volume would be needed when S_S and X_{S1} values were lower, as the utilization of X_{S2} might be required.

NUR tests gave valuable information on the preliminary estimation and design of BNR process. Together with the COD fractions the specific nitrate reaction rates can be used (Kujawa and Klapwijk, 1999) to calculate: (a) the theoretical anoxic zone volume required to denitrify the entire amount of nitrate-nitrogen present; (b) the maximum amount of nitrate-nitrogen concentration, which can be denitrified by the COD fraction present in the sewage, namely, Sewage Denitrification Potentials; and (c) the maximum influent-based nitrate-nitrogen, which can be denitrified in the existing anoxic zone of the activated sludge plant, namely, Plant Denitrification Potentials. Detailed calculations were carried out in this study, and it helped in the justification of the outcomes of the full- and laboratory-scale investigation (data not shown) (Cao et $al.$, 2003). Typical data included the maximum influent-based nitrate nitrogen concentration of 17 mg NO_3^--N l^{-1}, which can be denitrified in the existing anoxic zone of Phase IV of Bedok WRP during the peak flow period, and that of 32 mg NO_3^--N l^{-1}, which can be denitrified in the existing anoxic zone of Phase III of Kranji WRP during the base flow period (Cao et $al.$, 2003). The full- and laboratory-scale experimental data confirmed these predictions (Sections 3.3.5.1 and 4.3.4.1).

Table 2.4. Specific denitrification rates of Phase IV of Bedok WRP and Phase III of Kranji WRP (mg N (g VSS)$^{-1}$ h^{-1}).

Hydraulic flow condition	r_{DSS}		r_{DX1}		r_{DX2}		r_{Dend}	
	Bedok	Kranji	Bedok	Kranji	Bedok	Kranji	Bedok	Kranji
Peak	6.87	6.47	1.98	1.08	0.67	0.33	0.19	0.80
Peak (rainy)	8.82	n.a.	1.89	1.07	0.93	0.32	0.61	0.80
Low	8.55	n.a.	1.82	3.44	0.74	0.28	0.26	0.34
Normal	7.90	4.46	1.69	1.45	0.66	1.11	0.24	0.37
Average of the two plants	7.20		1.80		0.63		0.45	

2.3.5 Spatial distribution of denitrification activities

Figure 2.13 shows NO_3^--N uptake profiles of the NUR tests made with sludges taken from the anoxic zone, aerobic zones 1 and 3, and the RAS of the activated sludge process AU8 of Bedok WRP. The initial specific denitrification rates of the sludges taken from the respective locations were 7.8, 7.2, 6.6 and 8.5 mg NO_3^--N $(g\ VSS)^{-1}\ h^{-1}$, respectively. Compared with those in Table 2.4, these rates were predominantly governed by the utilization rate of the sewage S_S fraction. After approximately 30 minutes, the specific denitrification rates reduced to 4.7, 3.9, 3.5 and 2.7 mg NO_3^--N $(g\ VSS)^{-1}\ h^{-1}$, respectively, which were governed by the utilization rate of sewage X_{S1}. The specific denitrification rates of the sludges from the four different locations were comparable in both the initial and latter phases. This indicated that the adaptation of the microorganisms from the aerobic to the anoxic environment was rapid and the active denitrifier composition in the activated sludge of different compartments neared parity.

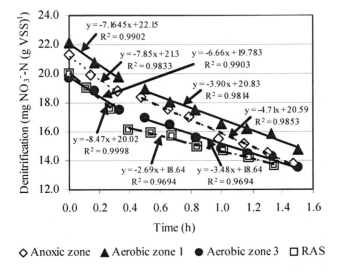

Figure 2.13. Denitrification rates of sludge sampled from the anoxic zone, aerobic zone 1 and zone 3, and the RAS of Bedok WRP.

2.3.6 Spatial distribution of nitrification activities

The NH_4^+-N uptake profiles of the AUR tests as shown in Figure 2.14 were made with sludges sampled from the anoxic zone, aerobic zones 1 and 3, and the RAS of the activated sludge process of AU8 of Bedok WRP. The respective specific nitrification rates were 7.5, 7.1, 6.4 and 7.3 mg NH_4^+-N $(g\ VSS)^{-1}h^{-1}$ with an average of 7.1 NH_4^+-N $(g\ VSS)^{-1}h^{-1}$. This was higher than 5.1 mg NH_4^+-N $(g\ VSS)^{-1}h^{-1}$, the average specific rate of the sludge taken from the pilot and full-scale activated sludge process under 20^0C in Denmark (Kristensen *et al.* 1992), but was close to the site value (Section 3.3.4). The differences between the rates were not significant indicating that the adaptation of microorganisms from the anoxic to the aerobic environment was rapid and the active nitrifier composition in the activated sludge of different compartments neared parity.

   The above mentioned results verified the rationale and practices of using similar kinetics (excluding yield) for nitrification and denitrification in both aerobic and anoxic compartments in modeling a single sludge activated sludge process.

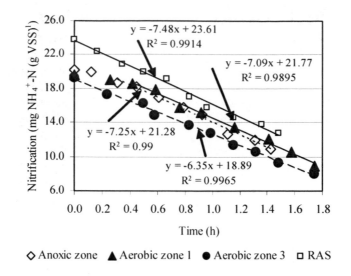

Figure 2.14. Nitrification rates of sludge sampled from the anoxic zone, aerobic zone 1 and zone 3, and the RAS of Bedok WRP.

2.4 SUMMARY

Diurnal hydraulic flow

For the Bedok, Kranji and Seletar WRPs, at least one peak flow with a factor of about 1.5 occurred in the day and a base flow with a factor ranging between 0.2 and 0.6 occurred between midnight and early morning. For some WRPs, another peak flow was observed in the evening. The COD and NH_4^+-N mass loading rates coincided with the respective first flow peak and base flow. The impact of supernatant recycling from the sludge dewatering operation was observed.

Conventional parameters

The weighted COD of the settled sewages of the three WRPs were 513, 526 and 366 mg COD l^{-1}, respectively, and all were within the 'concentrated' range. However, the SCOD of the settled sewages of the three WRPs were 100, 151 and 117 mg l^{-1}, respectively, which were within the 'diluted' range. The data indicated that the particulate COD of the settled sewages were high and, also, a higher inert particulate COD as indicated by the high COD/BOD_5 ratio.

The weighted TKN concentrations of the settled sewages of the three WRPs were 52.6, 50.0 and 43.0 mg N l^{-1}, which were within the 'moderate' range. The ammonia nitrogen concentrations were 37.4, 31.3 and 30.4 mg N l^{-1}, which were also within the 'moderate' range. The respective TP concentrations of the three WRPs were 13.7, 12.2 and 10.5 mg P l^{-1}, which were between the 'diluted' and 'moderate' ranges. The alkalinity values of the settled sewages of the three WRPs were 217, 179 and 148 mg l^{-1} (as $CaCO_3$), respectively, which were closer to the lower boundary of reported range. The ALK/NH_4^+-N ratio of 5.9 mg $CaCO_3$ (mg NH_4^+-N)$^{-1}$ was < 7.08 mg $CaCO_3$ (mg NH_4^+-N)$^{-1}$, and SCOD/TKN ratio was close to or < 2.86 mg COD (mg NO_3^--N)$^{-1}$.

Diluted SCOD, low alkalinity and low ratios of ALK/NH_4^+-N and SCOD/TKN indicated that (i) the soluble biodegradable COD might not be sufficient for efficient nutrient removal; (ii) either alkalinity addition or denitrification would be necessary when nitrification is facilitated; and (iii) a large anoxic reactor volume might be required for a high denitrification efficiency. The COD/BOD_5 ratios of the settled sewages of the three WRPs indicated a high portion ($\geq 20\%$) of inert solids in the settled sewages.

COD fractions

COD fingerprints, which were determined by NUR tests, varied with hydraulic flow. The readily biodegradable COD (S_S) of the settled sewage of Bedok WRP during the peak, normal and low flow periods were 66, 51 and 6 mg l^{-1} respectively, and the corresponding values were 36, 50 mg l^{-1} and non-detectable for that of Phase III of Kranji WRP. The S_S was almost depleted totally during the low flow due to the longer sewage retention time. The rapidly hydrolysable biodegradable COD (X_{S1}) of the settled sewage of Phase IV of Bedok WRP during the three flow periods were 60, 27 and 28 mg l^{-1} respectively, and the corresponding values were 81, 49 and 90 mg l^{-1} for Phase III of Kranji WRP. These data indicated that diurnal COD fractions should be used, rather than fixed COD fractions, in modeling and simulations.

Compared to the literature data, the S_S of the settled sewage investigated, which varied between non-detectable and 66 mg l^{-1}, was regarded as "very diluted" and lower than reported values. The elevated temperature of the warm climate could be the main reason.

Denitrification rates corresponding to COD fractions

The correlation between the magnitudes of the specific activities and the four corresponding COD fractions of the sludges from the Bedok and Kranji WRPs were logical and comparable. The average specific rate corresponding to S_S (r_{DSS}) was 7.20 mg N (g VSS)$^{-1}$ h^{-1}. The specific rate corresponding to X_{S1} (r_{DX1}) was 1.80 mg N (g VSS)$^{-1}$ h^{-1}. The specific rates corresponding to X_{S2} (r_{DX2}) and X_B (r_{Dend}) were 0.63 and 0.45 mg N (g VSS)$^{-1}$ h^{-1}, respectively.

Spatial distribution of denitrification and nitrification activities

The specific nitrification and denitrification activities of the activated sludges of the aerobic and anoxic compartments of Phase IV of Bedok WRP had no significant changes. This demonstrated that the shifting of environment under the site conditions did not affect the metabolisms of relevant microorganisms. This also demonstrated that the active nitrifier and denitrifier compositions in the sludges of the anoxic and aerobic compartments of a single sludge activated sludge process were comparable. This all verified the rationale to adopt the same set of kinetics of nitrification and denitrification in the aerobic and anoxic compartments of a single sludge activated sludge process with an appropriate anoxic growth factor in single sludge process modeling.

REFERENCES

American Public Health Association/ American Water Works Association/ Water Environment Federation (1998). Standard Methods for the Examination of Water and Wastewater. 20th ed., Washington D.C., USA.

Baumann B., Snozzi M., van der Meer J. R. and Zender A. J. B. (1997). Development of Stable Denitrifying Cultures during Repeated Aerobic –Anaerobic Transient Period. Wat. Res. **31**(8), 1947-1954.

Cao Yeshi (1994) Heterotrophic Biodegradation in Sewers and Drainage System: A Dual Phase Bioreactor, PhD dissertation, Delft University of Technology, Bakker & Balker Publ, Rotterdam, The Netherlands.

Cao Y. S., Ang C. M. and Zhao W. (2003) Performance Analysis of Full-and Laboratory-Scale Activated Sludge Process at Kranji and Bedok Water Reclamation Plants. Technical Report, Ref no. SUI/2001/030/TRTP1.

Cao Y. S., Ang C. M. and Raajeevan K. S. (2004a) Laboratory-Scale Studies of the Activated Sludge Process of Seletar Water Reclamation Plant. Technical Report, Ref no. SUI/2001/030/TRTP4.

Cao Y. S., Raajeevan K. S., Hu J. Y., Ang C. M., Teo K. H., Seah B. and Wah Y. L. (2005b) Characterization of Diurnal Settled Sewage and Spatial Nitrification and Denitrification Potentials of Activated Sludge of a Water Reclamation Plant in Singapore, 1st IWA-ASPIRE (Asia Pacific Regional Group) Conference & Exhibition, 10-15 July 2005, Singapore.

Çokgör E. B., Sözen S., Orhon D. and Henze M. (1998). Respirometric Analysis of Activated Sludge Behaviour-I: Assessment of the Readily Biodegradable Substrate. Wat. Res. **32**(2), 461-475.

Daigger G.T. and Nolasco D. (1995) Evaluation and Design of Full-Scale Wastewater Treatment Plants Using Biological Process Models. Wat. Sci. Tech. 31(2), 245-255.

Dold P.L., Ekama G. A. and Marais G. V. R. (1980). A General Model for the Activated Sludge Process. Prog. Wat. Tech. 12(6), 47-77.

Ekama G.A., Dold P.L. and Marais G. v. R. (1986). Procedures for Determining Influent COD Fractions and the Maximum Specific Growth Rate of Heterotrophs in Activated Sludge System. Wat. Sci. Tech., **18**, 91-114.

Federal Register, April 21, 1980, 45(78), 26811-268121, USA.

Gujer W., Henze M., Mino T. and van Loosdrecht M.C. M. (1999). Activated Sludge Model No. 3. Wat. Sci. Tech. **39**(1), 183-193.

Henze M., Grady C. P., Gujer W., Marais G. v. R. and Matsuo T. (1987). Activated Sludge Model No. 1. IAWPRC Sci. and Tech Report No. 1, IAWPRC, London.

Henze M. (1992) Characterisation of Wastewater for Modeling of Activated Sludge Process. Wat. Sci. Tech. **25**(6), 1-15.

Henze M., Gujer W., Mino T., Matsuo T., Wenzel M. C. and Marais G. v. R. (1995), Wastewater and Biomass Characterization for the Activated Sludge Model No. 2: Biological Phosphorus Removal. Wat. Sci. Tech. **31**(2), 13-23.

Henze M., Harremoës P., Janseen J. and Arvin E. (1997) Wastewater Treatment: Biological and Chemical Process, 2nd ed., Springer, Berlin.

Hydromantis Inc. (1999) Technical Reference of GPS.

Kappeler J. and Gujer W. (1992). Estimation of Kinetic Parameters of Heterotrophic Biomass under Aerobic Conditions and Characterization of Wastewater for Activated Sludge Modeling. Wat. Sci. Tech. **25**(6), 125-139.

Kristensen G. H., Jørgensen P. E. and Henze M. (1992). Characterization of Functional Microorganism Groups and Substrate in Activated Sludge and Wastewater by AUR , NUR and OUR. Wat. Sci. Tech. **25**(6), 43-57.

Kujawa K. and Klapwijk B.(1999). A Method to Estimate the Denitrification Potential for Pre-Denitrification Systems Using NUR Batch Test. Wat. Res. **33**(10), 2291-2300.

Langeveld J.G. (2004) Interactions Within Wastewater Systems, PhD dissertation, Delft University of Technology, The Netherlands.

Marais G.v.R. (1994) Wastewater Treatment: Activated Sludge Process. IHE, Delft, The Netherlands.

Metcalf and Eddy Inc. (2003) Wastewater Engineering: Treatment, Disposal and Reuse, 4th ed., McGraw-Hill, Washington, USA.

Naidoo V., Urbain V. and Burkley C. A. (1998). Characterization of Wastewater and Activated Sludge from European Municipal Wastewater Treatment Plants Using the NUR Test. Wat. Sci. Tech. **38**(1), 303-310.

Nelson P.E., Raunjkær K., Norsker N.H., Jenson N.A. and Hvitved-Jacobsen T. (1992) Transformation of Wastewater in Sewer System: A Review, Wat. Sci. Tech. **25**(6), pp: 17-31.

Raajeevan K.S. (2003) Biological Nitrogen Removal in Wastewater Treatment Plants in Singapore. M.Eng. Thesis. National University of Singapore.

Raunkjær K., Hvitved-Jacobsen T. and Nielsen P. (1995) Transformation of Organic Matter in a Gravity Sewer. Wat. Env. Res. **67**(2), 181-188.

Siegrist H and Tschui M. (1992). Interpretation of Experimental Data with Activated Sludge Model No. 1 and Calibration of the Model for Municipal Wastewater Treatment Plants. Wat. Sci. Tech. **25**(6), 167-183.

Spérandio M., Urbain V., Ginestet P., Audic M. J. and Paul E. (2001). Application of COD Fractionation by a New Combined Technique: Comparison of Various Wastewaters and Sources of Variability. Wat. Sci. Tech. **43**(1), 181-190.

Takàcs I. (2006) Personal Communication.

Vanrolleghem P. A., Spanjers H., Petersen B., Ginestet P. and Takacs I. (1999). Estimating (Combinations of) Activated Sludge Model No. 1 Parameters and Components by Respirometry. Wat. Sci. Tech. **39**(1), 195-214.

3

Performance of the full-scale activated sludge process

3.1 INTRODUCTION

Investigation and analysis of the performance of a full-scale activated sludge process is essential for improvement of the performance, operation, and optimal design of the system. Reports exist on the respective performance of different parts (zones) of the activated sludge process such as anoxic, aerobic compartments and RAS under the site's dynamic conditions (Concha and Henze, 1992; Lesouef et al., 1992; Siegrist and Tschui, 1992; Concha and Henze, 1996; van Veldhuizen et al., 1999; Harrbo et al., 2001; Meijer, 2004); but, in general, the information is still limited. Constraints on the amount of work involved might be the major reason. This affects our understanding of the process and the performance of the full-scale activated sludge process, and the applicability of models in simulating 'real' systems. As already mentioned, little practical information is available on BNR in the full-scale activated sludge process in warm climates.

The objectives of the investigation of the performance of the full-scale activated sludge process are defined as follows:

i   to obtain a clear picture of the performance of the full-scale activated sludge process, mainly the modified Ludzack-Ettinger (MLE) process including the performance of the anoxic and aerobic compartments and the secondary clarifier under the site's diurnal conditions. These data enables the construction of a consolidated mass flow and balance, an assessment of the whole system performance in meeting the feed requirements of the NEWater plants and the evaluation of the design of the upgraded activated sludge process;

ii   to collect the key design parameters of the activated sludge process and the operation parameters that are adopted in the design of scaled-down laboratory experiments. Then, the results of the laboratory-scale experiment can be compared with those of the full-scale process to justify the feasibility of the scale-down principle developed for the activated sludge process; and

iii   to collect sufficient data and information for verifying the parameters calibrated using data of the laboratory-scale process in modeling and simulation. This helps in studying the feasibility to model the performance and design of the full-scale process by using the laboratory-scale process data.

Full-scale activated sludge processes of Bedok and Kranji WRPs were selected for the detailed investigation in this study (Cao *et al.*, 2003). The results of the activated sludge process of Phase IV of Bedok WRP are presented in this chapter, while part of the results of the investigation of Kranji WRP are adopted for parameter verification, as presented in Section 5.4.3.

3.2 MATERIALS AND METHODS

3.2.1 Site conditions and sampling programme

Phase IV of Bedok WRP. There are eight similar activated sludge process trains in Bedok WRP. There are four Phases with two trains in each phase. The effective volume of each tank (train) is 8,724 m^3 with dimensions of 87 m x 21.8 m x 4.6 m (length x width x depth). For each train, there are four compartments of equal size, but only the first compartment has a physical partition. Phase IV, consisting of activated sludge trains 7 (AU7) and 8 (AU8), which share common secondary sedimentation tanks, was retrofitted for the NEWater production, and was selected in this study. Of these two trains, one is an MLE and the other is an LE activated sludge process. An anoxic zone at the head of each activated sludge tank occupies a volumetric ratio of 25%. A surface aerator was used to provide air supply in the each aeration compartment of each train. Figure 3.1 shows the process configuration and key parameters.

Sampling programme. A 24-h sampling and monitoring programme was implemented between 22 and 23 April 2003 at Bedok WRP. Generally, April is not a rainy month in Singapore, so the sampling was under dry weather conditions. The locations of the sampling points (SP) together with the key design parameters are shown in Figure 3.1.

The sampling parameters included diurnal flow and conventional parameters such as COD, SCOD, BOD_5, TSS, TN, TKN, NH_4^+-N, NO_3^--N, DO, pH and alkalinity, etc. The sampling points were the inlets of the activated sludge process, individual anoxic and aerobic compartments and return activated sludge (RAS) with a bi-hourly sampling frequency.

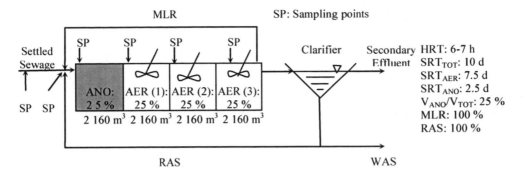

Figure 3.1. Schematic diagram of the MLE activated sludge process of Phase IV of Bedok WRP and sampling points in the monitoring programme.

3.2.2 Analysis

The samples were analyzed for those parameters mentioned above. The samples from the anoxic and aerobic compartments and the RAS were filtered on site immediately after sampling to avoid possible reactions in the sampling bottles. These samples were analyzed for pH, SCOD, NH_4^+-N, NO_3^--N and alkalinity. Mixed liquor samples were also sampled from the respective compartments and the RAS for analysis of MLVSS. The temperature and DO concentration of the anoxic and aerobic compartments and the RAS were recorded during sampling. Analytical methods were similar to those described in Chapter 2.

3.2.3 Calculations

Mass balance is constructed across the control volume (compartment) using the equation below. The term C is the concentration of a chemical i.e., NH_4^+-N and NO_3^--N etc. $V(\Delta C/\Delta t)$ is the accumulation term; and under steady state, $V(\Delta C/\Delta t)$ is zero. Δt can be 1 or 2 h depending upon sampling frequency. The mass formation or removal rate $(\Delta M/\Delta t)_{f/r}$ due to various biochemical reactions taking place in the control volume can be calculated when a set of hydraulic flow and concentration profile data is available.

$$V\frac{\Delta C}{\Delta t} = Q_{in}C_{in} - Q_{out}C_{out} + \left(\frac{\Delta M}{\Delta t}\right)_{f/r,t}$$

(3.1)

The influent-based formation or removal concentration in the control volume during a time interval, $C_{f/r,t}$, is expressed in the equation below.

$$C_{f/r,t} = \frac{1}{Q_{in,t}}\left(\frac{\Delta M}{\Delta t}\right)_{f/r,t}$$

(3.2)

3.3 RESULTS AND DISCUSSION

Only the results obtained from the MLE process were introduced in the following sections although the LE activated sludge process, which shared the common sewage and secondary clarifier with the MLE process, was also studied in detail (Cao *et al.*, 2003).

3.3.1 Key design and operation parameters

Given the uncertainty due to grab samples in modeling (Meijer, 2004), information on design and operation parameters was collected through (a) the design manual; and (b) study of the plants' record on influent mass, sludge wasting, MLSS etc. One typical parameter was the SRT, which was checked though calculation based on the averages of COD mass load, MLSS and wasting rate. The key design parameters of the activated sludge processes of Phase IV of Bedok WRP are: $SRT_{TOT} = 10$ d, $SRT_{ANO} = 2.5$ d, calculated based on an anoxic volumetric ratio of 25% and assuming similar MLSS concentration in the activated sludge compartments, $SRT_{AER} = 7.5$ d, HRT = 6-7 h, MLSS $\approx 2\,800$ mg l^{-1}, MLR ratio of 100% for the MLE process, RAS = 100% for both trains and operating temperature = $30 \pm 1°C$. The hydraulic flow pattern of the activated sludge process is approximately a plug flow based on the ratio of the length over the width, and longitudinal mixing could occur (Burrow *et al.*, 2001). These parameters, together with the site influent characterization data, were used in the design of the laboratory-scale activated sludge system. The reasons for choosing these parameters as the 'ruling regime' of the full-scale activated sludge process of Bedok WRP are elaborated in Section 4.1.1.

3.3.2 Influent characterization

3.3.2.1 Diurnal hydraulic flow and carbonaceous and nitrogenous mass loadings

Figure 3.2 shows the diurnal hydraulic flow of a single train, which was calculated based on the pumping station flow record divided by the number of the trains (8). Differing from the flow pattern of Figure 2.1, two peaks were recorded. The first hydraulic peak flow lasted between 08:00 and 14:00 while the second between 19:30 and 22:00 with peak factors of about 1.2 for both. However, the duration of the second peak flow was shorter than the first. The low flow occurred between 02:00 and 06:00 with a flow factor of about 0.5. The actual HRT of the day of sampling was 7.3 h.

The COD and NH_4^+-N mass loading rates were calculated based on the diurnal influent (Figure 3.2) and the COD and NH_4^+-N concentration profiles (Figures 3.4 and 3.5). As can be seen in Figure 3.2, generally, the COD mass loading rate peaks coincided with the hydraulic flow peaks, and the mass loading rate of the second peak was about 15% higher than that of the first. Figure 3.3 shows that the NH_4^+-N mass loading rate flow peaks also occurred during the daytime and coincided with the peak flows, although the second one was lower than the first. No peak was recorded in the early morning because only one dewatering centrifuge unit was in operation during the sampling period as opposed to two during normal operation.

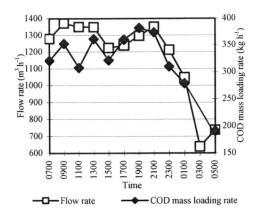
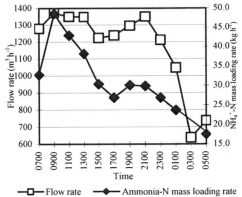

Figure 3.2. Diurnal flow and COD mass loading rate profiles of each train of Phase IV of Bedok WRP (22–23 April 2003).

Figure 3.3. Diurnal flow and NH_4^+-N mass loading rate profiles of each train of Phase IV of Bedok WRP (22–23 April 2003).

3.3.2.2 Conventional parameters

As shown in Figure 3.4, the COD, BOD_5 and TSS concentration peaks, except SCOD, coincided, to different extents, with the hydraulic flow peaks and, most likely, due to a re-suspension of the settled TSS. Compared to the concentration profiles in Figure 2.5, which was obtained between 27 and 28 December 2002, the average COD here was about 200 mg l^{-1} less due to the low TSS concentrations.

For the concentration profiles of nutrient constituents including TKN, NH_4^+-N and TP (Figure 3.5), the concentration peaks of TKN coincided with the hydraulic flow peaks and to a lesser extent, the concentration of TP. The reason might be similar to that of the increase in COD, BOD_5 and TSS. The ammonia nitrogen concentration increased only in the morning peak flow period.

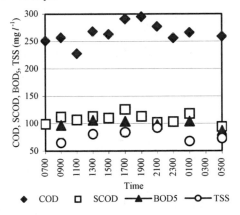
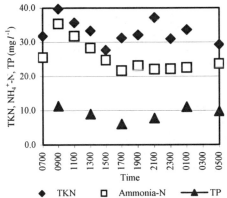

Figure 3.4. Diurnal COD, SCOD, BOD_5 and TSS concentration profiles of the settled sewage of Phase IV of Bedok WRP (22–23 April 2003).

Figure 3.5. Diurnal TKN, NH_4^+-N and TP concentration profiles of the settled sewage of Phase IV of Bedok WRP (22–23 April 2003).

3.3.3 Carbonaceous matter removal

Carbonaceous matter removal was studied mainly through SCOD reduction. Figure 3.6 shows the diurnal SCOD profiles of the feed settled sewage and in each of the four compartments. Two SCOD peaks in the anoxic reactor occurred and then gradually dampened in the first, second and third aerobic compartments. The average SCOD values of the settled sewage and four compartments were 109, 24, 14, 18 and 11 mg l^{-1} respectively, illustrating that most of the SCOD was removed in the anoxic compartment due to denitrification and in the first aerobic compartment through carbon oxidation, while SCOD removal in the second and third aerobic compartments was low. The average SCOD removal efficiency, which was calculated by the equation below, was 90.0%.

$$SCOD \text{ removal efficiency } (\%) = \frac{[SCOD]_{INF} - [SCOD]_{EFF}}{[SCOD]_{INF}} \times 100 \qquad (3.3)$$

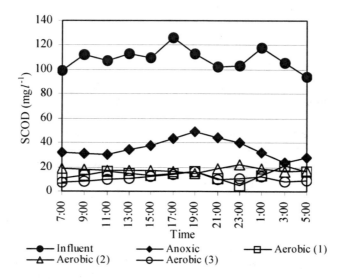

Figure 3.6. SCOD concentration profiles of the influent and the four compartments of the MLE activated sludge process of Phase IV of Bedok WRP.

3.3.4 Nitrification

Figure 3.7 shows the diurnal NH$_4^+$-N concentration profiles of the settled sewage feed and of the four compartments. Two NH$_4^+$-N concentration peaks occurred in the anoxic compartment. The first occurred between 09:00 and 14:00 and corresponded to the first NH$_4^+$-N mass loading rate peak. The second, which was lower than the first, occurred between 18:00 and 21:00 and corresponded to the second hydraulic peak. The peaks were gradually dampened in the first and second aerobic compartments and almost disappeared in the third aerobic compartment, demonstrating the effect of SRT on the fluctuations in effluent quality.

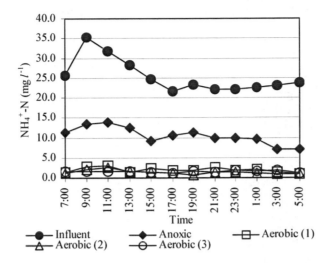

Figure 3.7. NH_4^+-N concentration profiles of the influent and the four compartments of the MLE activated sludge process of Phase IV of Bedok WRP.

The average diurnal NH_4^+-N concentrations of the settled sewage and four compartments were 25.6, 10.3, 2.0, 1.4 and 1.6 mg NH_4^+-N l^{-1}, respectively, indicating that the NH_4^+-N concentration profiles of the three aerobic compartments were close to each other. The ammonia - nitrogen concentration reduction corresponded to the areas enclosed within the NH_4^+-N concentration curves of the anoxic and each of the three aerobic compartments, as shown in Figure 3.7. The diurnal NH_4^+-N concentrations of the first aerobic compartment were << 5 mg NH_4^+-N l^{-1}, the feed quality requirement of the NEWater plant. Most of the NH_4^+-N in the feed was removed in the first aerobic compartment and the contributions of the second and third compartments in NH_4^+-N oxidation were marginal.

The specific nitrification rate (SNR) was obtained through the NH_4^+-N mass removal rate, which was obtained through dividing the NH_4^+-N mass oxidation rate (calculated using Equation 3.1) by (MLVSS x V_{AER}). Figures 3.8(a), (b), and (c) show the variation of the specific nitrification rate in the three aerobic compartments respectively, which indicate that the response of the nitrifiers to ambient NH_4^+-N mass load was almost instantaneous. This capability of the nitrifiers attenuated the effect of feed NH_4^+-N mass load shock on effluent quality. Also, it showed that under dynamic conditions, the NH_4^+-N mass load is another factor governing the nitrifiers activity apart from the aerobic SRT under site conditions.

The calculated specific nitrification rates shown in Table 3.1 illustrated an obvious correlation between the specific activity and the NH_4^+-N-based F/M ratio. The average specific rate in the first aerobic reactor was 8.7 g NH_4^+-N (kg VSS)$^{-1}$ h^{-1}, which was close to 7.1 g NH_4^+-N (kg VSS)$^{-1}$, the average value obtained from the AUR test (Section 2.3.6).

The NH_4^+-N-based diurnal nitrification efficiencies of the first aerobic compartment and the overall activated sludge process, which were calculated by the equation below, are presented in Figure 3.9. The average NH_4^+-N removal efficiency in the first aerobic compartment was 92.2%. The total nitrogen-based nitrification efficiencies are presented in Section 3.3.8.2.

$$\text{Nitrification efficiency } (NH_4-N \text{ based}) (\%) = \frac{[NH_4^+-N]_{INF} - [NH_4^+-N]_{EFF}}{[NH_4^+-N]_{INF}} \times 100 \qquad (3.4)$$

Table 3.1. Specific nitrification rates and F/M ratios of the three aerobic compartments of Phase IV of Bedok WRP.

Compartment	SNR range [g NH₄⁺-N (kg VSS)⁻¹ h⁻¹]	SNR_avg [g NH₄⁺-N (kg VSS)⁻¹ h⁻¹]	F/M range [kg NH₄⁺-N h⁻¹ (kg VSS)⁻¹ d⁻¹]	(F/M)_avg [kg NH₄⁺-N h⁻¹ (kg VSS)⁻¹ d⁻¹]
AER 1	5.6 - 12.0	8.7	0.17 - 0.36	0.26
AER 2	0.1 - 1.2	0.7	0.02 - 0.06	0.05
AER 3	0.0 - 0.7	0.3	0.02 - 0.07	0.04

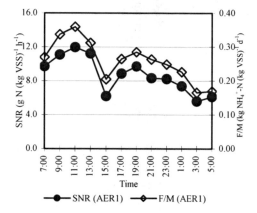

Figure 3.8(a). Variations of the specific nitrification rate with the NH₄⁺-N-based F/M ratio in the first aerobic compartment of the MLE activated sludge process of Phase IV of Bedok WRP.

Figure 3.8(b). Variations of the specific nitrification rate with the NH₄⁺-N-based F/M ratio in the second aerobic compartment of the MLE activated sludge process of Phase IV of Bedok WRP.

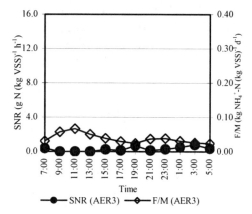

Figure 3.8(c). Variations of the specific nitrification rate with the NH₄⁺-N-based F/M ratio in the third aerobic compartment of the MLE activated sludge process of Phase IV of Bedok WRP.

Figure 3.10 shows the average diurnal DO, SCOD and NH_4^+-N concentration profiles of the four compartments. A large portion of the SCOD and most of the NH_4^+-N were removed in the first aerobic compartment and, as a result, the DO concentration in the first aerobic compartment was relatively low compared with the other two aerobic compartments. The third aerobic compartment seemed to contribute little to SCOD and NH_4^+-N removals. These site data illustrated the possibility that the aerobic SRT of 7.5 d might be shortened and the volume of aerobic compartments in the existing activated sludge process (75% of total volume) could be reduced. The relevant studies were performed in the laboratory (Section 4.3) and modeling (Sections 5.5.1.6 and 5.5.2).

Almost similar COD removal and nitrification efficiencies were also found in the LE activated sludge process showing that MLR had no significant effect, as expected, on COD and NH_4^+-N removals under site conditions.

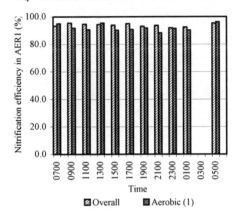 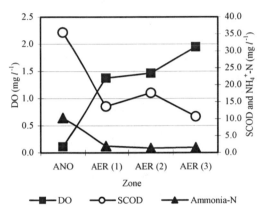

Figure 3.9. Diurnal nitrification efficiencies of the first aerobic compartment with respect to the overall nitrification efficiency of the MLE activated sludge process of Phase IV of Bedok WRP.

Figure 3.10. Average DO, SCOD and NH_4^+-N concentration profiles of the four compartments of the MLE activated sludge process of Phase IV of Bedok WRP.

3.3.5 Denitrification

3.3.5.1 Denitrification in the activated sludge tanks

Figure 3.11 shows the diurnal NO_3^--N concentration profiles of each of the four compartments. The average NO_3^--N concentrations were 1.2, 6.9, 8.2 and 8.9 mg NO_3^--N l^{-1}, respectively. In the anoxic compartment, the nitrate-nitrogen concentration was < 1.0 mg NO_3^--N l^{-1} between 07:00 and 13:00 and between 01:00 and 07:00. The NO_3^--N concentration was between 2 and 3 mg NO_3^--N l^{-1} between 14:00 and 22:00, and no appropriate reasons were found to explain this as yet.

Significant increases in NO_3^--N concentration occurred in the first aerobic compartment corresponding to NH_4^+-N oxidation. However, in the second and third aerobic compartments, the NO_3^--N concentration increases during the morning peak period were in the range between 0.7 and 1.4 mg NO_3^--N l^{-1} only since NH_4^+-N reduction was modest in these two compartments.

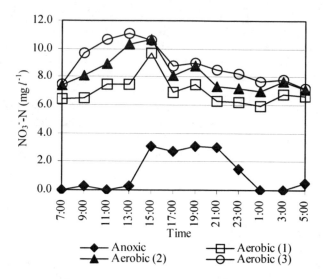

Figure 3.11. NO$_3^-$-N concentration profiles of the four compartments of the MLE activated sludge process of Phase IV of Bedok WRP.

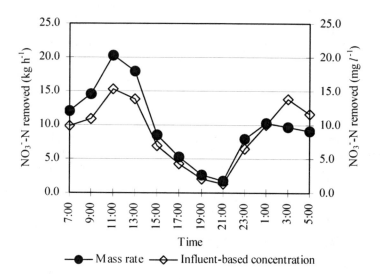

Figure 3.12. NO$_3^-$-N removed, in terms of the mass removal rate and influent-based concentration, in the anoxic compartment of the MLE activated sludge process of Phase IV of Bedok WRP.

Figure 3.12 shows the NO$_3^-$-N mass removal rate and influent-based NO$_3^-$-N removal in the anoxic compartment, which were calculated using equations3.1 and 3.2. The average influent-based nitrate-nitrogen concentration removal was 8.9 mg NO$_3^-$-N l^{-1} although variations in diurnal nitrate nitrogen removal were apparent. The maximum influent-based nitrate-nitrogen concentration removal was 15.1 mg NO$_3^-$-N l^{-1} at 11:00 during the morning peak flow period, which was very close to 17 mg NO$_3^-$-N l^{-1}, the sewage maximum denitrification potential during the peak flow period predicted by sewage COD fractionation study (Cao *et al.*, 2003).

The calculated diurnal specific denitrification rates in the anoxic compartment, which were obtained through dividing the NO_3^--N mass removal rate (calculated using Equation 3.1) by (MLVSS x V_{ANO}), are shown in Figure 3.13. The activity varied between 0.4 and 4.2 g NO_3^--N (kg VSS)$^{-1}$ h^{-1} with an average of 2.1 g NO_3^--N (kg VSS)$^{-1}$ h^{-1}. The highest value of 4.2 g NO_3^--N (kg VSS)$^{-1}$ h^{-1} was < 7.2 g NO_3^--N (kg VSS)$^{-1}$ h^{-1}, the specific rate corresponding to the soluble readily biodegradable COD (S_S) but higher than 1.8 g NO_3^--N (kg VSS)$^{-1}$ h^{-1}, the denitrification rate corresponding to the rapidly hydrolysable biodegradable COD (X_{S1}). The lowest value of 0.4 g NO_3^--N (kg VSS)$^{-1}$ h^{-1} was comparable with the average specific denitrification rate of 0.63 mg NO_3^--N l^{-1} corresponding to slowly hydrolysable biodegradable COD (X_{S2}) (Table 2.4).

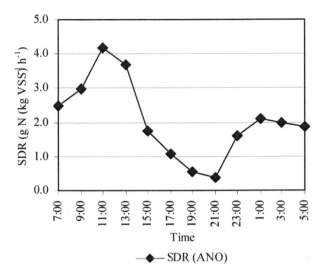

Figure 3.13. Specific denitrification rates in the anoxic compartment of the MLE activated sludge process of Phase IV of Bedok WRP.

3.3.5.2 Denitrification in the final clarifier

Figure 3.14 shows that the NO_3^--N concentrations at the inlet of the clarifier with an average of 8.9 mg NO_3^--N l^{-1} were higher than those at the outlet of the clarifier, and that the NO_3^--N concentrations of the RAS with an average of 3.4 mg NO_3^--N l^{-1} were the lowest. These profiles and figures indicated that denitrification had occurred in the sludge blanket of the FC. The NO_3^--N removals in the FC, calculated using the same equations as those for anoxic compartment, are shown in Figure 3.15. The maximum nitrate-nitrogen mass removal rate reached, in the evening, was 10.3 kg h^{-1}, which was about half of the value of the anoxic compartment. The average influent-based NO_3^--N concentration removal was 6.8 mg NO_3^--N l^{-1}, which was close to that of the anoxic compartment. The average value of 8.9 mg NO_3^--N l^{-1} at the inlet of the FC was close to 10 mg NO_3^--N l^{-1}, the limit for the design of the FC to prevent sludge overflow (Henze *et al.*, 1993; Ekama *et al.*, 1997), and should therefore be further reduced.

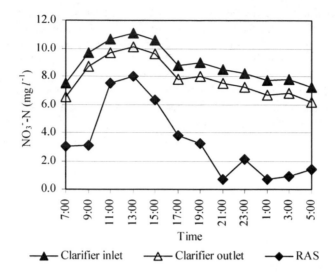

Figure 3.14. NO_3^--N concentration profiles at the inlet and outlet of the clarifier and of the RAS of the MLE activated sludge process of Phase IV of Bedok WRP.

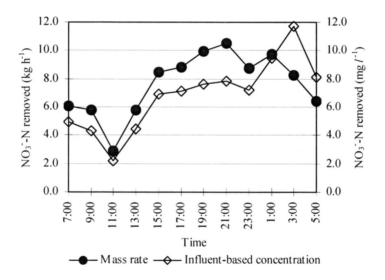

Figure 3.15. NO_3^--N removed, in terms of the mass removal rate and influent-based concentration, in the clarifier of the MLE activated sludge process of Phase IV of Bedok WRP.

3.3.5.3 Denitrification efficiency

The overall denitrification in the activated sludge process was contributed by the anoxic compartment, the FC and the aerobic compartments through simultaneous nitrification and denitrification (SND). The denitrification efficiencies of the anoxic compartment and the FC were calculated based on NO_3^--N produced in the whole process and denitrified in the anoxic compartment and the FC. Assuming organic nitrogen (TKN-NH_4^+-N) was used for assimilation, the NH_4^+-N-based denitrification efficiencies of the anoxic compartment and FC were calculated by the equations below.

$$\text{Denitrification efficiency (\%)} = \frac{Q_{IN} \times [NO_3^- - N]_{IN} - Q_{OUT} \times [NO_3^- - N]_{OUT}}{Q_{INF} \times \{[NH_4^+ - N]_{INF} - [NH_4^+ - N]_{EFF}\}} \times 100 \qquad (3.5)$$

where $[NH_4^+\text{-N}]_{INF} - [NH_4^+\text{-N}]_{EFF} \approx [NO_3^-\text{-N}]$

The overall denitrification efficency was calculated by the equation as follow:

$$\text{Overall denitrification efficiency (\%)} = \{1 - \frac{[NO_3^- - N]_{EFF}}{[NH_4^+ - N]_{INF} - [NH_4^+ - N]_{EFF}}\} \times 100 \qquad (3.6)$$

The contribution of the SND was calculated by that of the difference between the denitrification efficiency of the overall process and that of denitrification of the anoxic compartment and the FC. Figure 3.16 shows the overall denitrification efficiency and the contributions of each component. The average denitrification efficiencies of the anoxic compartment, FC and aerobic compartments were 34.0%, 23.0% and 5.1% respectively. The average diurnal denitrification efficiency of the whole process was 62.1%. The total nitrogen-based denitrification efficiencies are presented in Section 3.3.8.

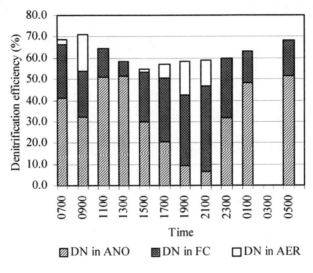

Figure 3.16. Diurnal denitrification efficiencies of the various functional zones of the MLE activated sludge process of Phase IV of Bedok WRP.

3.3.6 pH and alkalinity

The diurnal pH data in the anoxic and three aerobic compartments exhibited a reduced tendency in sequence as shown in Figure 3.17. The diurnal average values were 6.56, 6.32, and 6.25 in the three aerobic compartments, respectively. Corresponding to denitrification and nitrification, the alkalinity data exhibited a similar reduced tendency as shown in Figure 3.18. The average alkalinities were 92.6 mg in the anoxic zone, 47.9 mg, 41.1 mg and 37.0 mg as $CaCO_3$ l^{-1} in the first, second and last aerobic compartments, respectively, where the average alkalinity of the settled sewage was 172.8 mg as $CaCO_3$ l^{-1}.

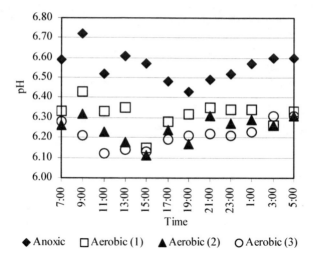

Figure 3.17. pH profiles in the four compartments of the MLE activated sludge process of Phase IV of Bedok WRP.

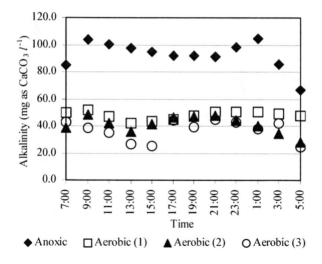

Figure 3.18. Alkalinity profiles in the four compartments of the MLE activated sludge process of Phase IV of Bedok WRP.

3.3.7 Nitrogen in the final effluent

Figure 3.19 shows the concentrations of soluble nitrogen components in the effluent. The average diurnal soluble TN concentration was 10.4 mg N l^{-1}, and NO_2^--N was not included although it ranged between 0.1 and 0.8 mg NO_2^--N l^{-1}. The average diurnal ammonia-nitrogen concentration was 1.6 mg NH_4^+-N l^{-1}, which was « 5 mg NH_4^+-N l^{-1}, the feed water quality requirement of the NEWater plants. The overall nitrification efficiency calculated by Equation 3.4 was 93.8%. The average organic ammonia-nitrogen (soluble TKN-NH_4^+-N) was only 0.2 mg NH_4^+-N l^{-1}.

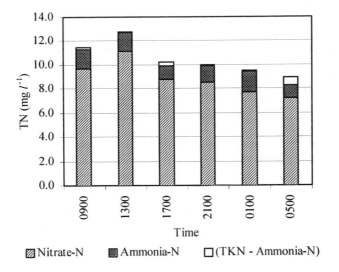

Figure 3.19. Soluble nitrogen components concentrations in the effluent of the third aerobic compartment of the full-scale MLE activated sludge process of Phase IV of Bedok WRP.

Comparing the nitrogen removal of the MLE and LE activated sludge trains, the general trend was that denitrification in the anoxic compartment was higher due to the supply of NO_3^--N by MLR compared to that of the LE activated sludge train. The denitrification in the anoxic compartment of the MLE activated sludge train was about 6 mg NO_3^--N l^{-1} higher than that of the LE activated sludge train. The specific denitrification rate in the anoxic compartment of the MLE activated sludge train was almost double that of the anoxic compartment of the LE activated sludge train (Cao *et al.*, 2003). The average overall denitrification efficiency of the MLE process, calculated by Equation 3.6, was 62.1%, which was about 10% higher than that of the LE train.

3.3.8 Mass balance and yield coefficients

3.3.8.1 COD balance

A COD mass balance exemplifies the overall picture depicting COD conversion and mass flow in the activated sludge process. The balance was constructed by accounting for all forms of COD input into, conversion within and output from the system. The influent and effluent COD was calculated based on the diurnal flow and SCOD and COD profiles. Others were COD removals in the anoxic compartment and secondary clarifier (due to denitrification), RO_H, oxygen requirement (consumption) in the aerobic compartments due to carbonaceous matter oxidation, anabolic COD (assimilation into biomass, equivalent to waste sludge COD), and soluble and particulate COD in the final effluent. Waste sludge COD was taken as 10% of the COD content of the VSS in the reactor compartments based on an SRT of 10 d. The COD catabolized in the anoxic and final clarifier were calculated based on the COD/NO_3^--N ratio of 2.86 (Henze *et al.*, 1997). Oxygen requirement in the aerobic compartments due to carbonaceous matter oxidation, RO_H, was calculated as the difference between the influent COD and the sum of the other compartments (Grady *et al.*, 1999). Table 3.2 summarizes the data calculated on a 24-h basis.

Table 3.2. Distribution of the COD (kg COD d^{-1}) in the MLE activated sludge process of Phase IV of Bedok WRP.

Influent COD	Denitrified COD in anoxic compartment	Denitrified COD in final clarifier	Aeration Removed COD (RO$_H$)	Anabolized COD$_{WAS}$	COD in final effluent	
					Soluble	Particulate
7 434	691	525	3 506	2 047	159	506

Figure 3.20 shows the percentage distribution of the influent COD in the MLE activated sludge process of Phase IV of Bedok WRP, in descending order, based on the values of Table 3.2: 47%, oxygen-COD consumption in the aerobic compartments; 28%, COD assimilation into biomass; 9%, COD removal in the anoxic compartment; 7%, COD removal in the final clarifier (i.e. 16% COD removal in total in the anoxic compartment and the final clarifier); 7%, particulate COD in the final effluent; and 2% soluble COD in the final effluent. The balance shows that oxygen consumption in the aerobic compartments was higher than COD removed in the anoxic compartment.

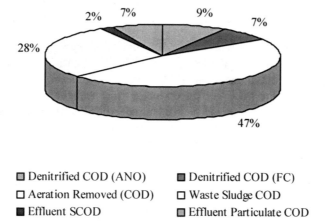

Figure 3.20. Percentage distribution of the influent COD in the MLE activated sludge process of Phase IV of Bedok WRP.

Based on the COD balance data (Table 3.2), the net sludge yield coefficients (COD of anabolism) (Grady *et al.*, 1999) and COD catabolism can be calculated from the two equations below, respectively.

$$\text{Net sludge yield }(\%) = \frac{COD_{WAS}}{COD_{INF} - COD_{EFF}} \times 100 \tag{3.7}$$

$$\text{COD catabolized }(\%) = \frac{COD_{DN} + RO_H}{COD_{INF} - COD_{EFF}} \times 100 \tag{3.8}$$

The net sludge (anabolism) yield calculated was 30.2%, which was very close to the statistical data in Singapore and the design data under the same temperature conditions (WEF, 1992). The calculated total COD catabolism coefficient in the system was 69.8%. Given the data in Table 3.2, the process oxygen coefficient, which was calculated by dividing RO_H by COD removed, was $0.52\,g$ O_2 (g $COD_{REM})^{-1}$. Comparing $0.52\,g\,O_2\,(g\,COD_{REM})^{-1}$ with a COD catabolism coefficient of 69.8%, a 26% [(69.8-52)/69.8] reduction on carbonaceous oxygen requirement due to denitrification in the anoxic zone and FC was achieved. Further, taking the oxygen requirement for autotrophic ammonia oxidation, RO_A, into account – calculated by [4.57 x (TKN_{IN}-N_{ASSIM}-TKN_{EFF}] (Grady *et al.*, 1999) and the data in Table 3.3 – a 12% total oxygen demand reduction [26% x (RO_H/(RO_H+RO_A))], was achieved.

3.3.8.2 Nitrogen balance

The approaches in the calculation of the COD mass balance were similarly adopted for the calculation of the nitrogen mass balance. Table 3.3 summarizes the nitrogen mass balance data on a 24-h basis. The total nitrogen input into the system was 929 kg N, which was the sum of the total nitrogen removed in the anoxic reactor and final clarifier, that assimilated into cells (calculated based on a SRT of 10 d and 10% nitrogen content of the VSS in wasting) and that discharged in the effluent (calculated from the TSS in the effluent and 10% nitrogen content of the VSS). This total nitrogen input was nearly the same as the influent TKN load into the system.

Table 3.3. Distribution of the nitrogen (kg N d^{-1}) in the MLE activated sludge process of Phase IV of Bedok WRP.

Nitrogen removed in anoxic compartment	Nitrogen removed in final clarifier	Nitrogen in wasted sludge (WAS)	Nitrogen in final effluent				Calculated influent TN (TKN)
			NH_4^+-N	NO_3^--N	Sol. Org-N	Susp. TKN	
241	183	170	44	222	28	41	929

Figure 3.21 shows the percentage distribution of input nitrogen in the activated sludge process. The percentage distribution in descending order was: 26%, NO_3^--N removal in the anoxic compartment; 24%, NO_3^--N in the final effluent; 20%, NO_3^--N removal in the final clarifier; 18%, nitrogen assimilation into sludge cells; 5%, NH_4^+-N in the final effluent; 4%, suspended TKN in the final effluent; and 3%, soluble organic nitrogen in the final effluent.

Differing from the equations in sections 3.3.4 and 3.3.5.3, where calculations were based on ammonia-nitrogen while assimilation was unaccounted for, the nitrification and denitrification efficiencies can be calculated based on the nitrogen balance and taking nitrogen assimilation into account with the two equations below:

$$\text{Nitrification efficiency (\%)} = \frac{TN_{IN} - TKN_{OUT} - N_{ASSIM}}{TN_{IN} - N_{ASSIM}} \times 100 \tag{3.9}$$

$$\text{Denitrification efficiency (\%)} = \frac{N_{DN}}{TN_{IN} - TKN_{OUT} - N_{ASSIM}} \times 100 \tag{3.10}$$

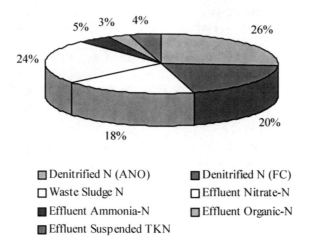

Denitrified N (ANO)   Denitrified N (FC)
Waste Sludge N   Effluent Nitrate-N
Effluent Ammonia-N   Effluent Organic-N
Effluent Suspended TKN

Figure 3.21. Percentage distribution of the influent nitrogen in the MLE activated sludge process of Phase IV of Bedok WRP.

The calculated nitrification efficiency was 85.1%, and the overall denitrification efficiency calculated was 65.6%, of which 37.3% was attributed to the anoxic compartment and 28.3% to the final clarifier. Table 3.4 records the nitrification and denitrification efficiencies calculated from both the ammonia-nitrogen and total nitrogen.

Table 3.4. Nitrification and denitrification efficiencies (%) of the MLE activated sludge process of Phase IV of Bedok WRP.

Nitrification		Denitrification in the anoxic zone and clarifier (based on total nitrogen)		Total denitrification	
Based on NH_4^+-N	Based on total nitrogen	Anoxic zone	Final clarifier	Based on NH_4^+-N	Based on total nitrogen
93.8	85.1	37.3	28.3	62.1	65.6

3.3.9 Efficiency of aeration

The aeration energy constitutes one of the major operating costs of a municipal wastewater treatment plant. However, limited data is available on the aeration efficiency under warm climate conditions. The aeration efficiency shown in Table 3.5 was calculated using the two equations below. The calculation of RO_H and RO_A were introduced in the last section, and the aeration energy supplied was calculated according to the number, types and the power ratings of the surface aerators, and assuming 1.83 kg O_2 $(kWh)^{-1}$ as the oxygen transfer rate of the surface aerators (Metcalf and Eddy, 2003). The calculated aeration efficiency was 37.9%, which was in the high range as reported in literature.

$$\sum Oxygen\ Consumption = RO_H + RO_A \qquad (3.11)$$

$$Aeration\ Efficiency\ (\%) = \frac{\sum Oxygen\ consumption}{\sum Aeration\ energy\ supplied} \times \frac{kWh}{1.83\ kg\ O_2} \times 100 \qquad (3.12)$$

Table 3.5. Aeration efficiency (%) of the MLE activated sludge process of Phase IV of Bedok WRP and data used in the calculation.

Total aeration energy supplied (kWh d^{-1})	Total oxygen consumption due to COD and NH$_4^+$-N removals (kg d^{-1})	Aeration efficiency
9 312	6 458	37.9

3.4 SUMMARY

Performance of the full-scale activated sludge process

Key parameters. The key design parameters including the configuration, sizes, SRT, HRT, and ratios of MLR and RAS etc., of the MLE activated sludge process of Bedok WRP were collected through (i) the design manual; and (ii) study of the plants' record. The SRT was verified through the calculation based on the statistical data of COD mass load, MLSS and wasting rate.

Influent characterization. Diurnal hydraulic flow, carbonaceous and nitrogenous mass loads, and conventional parameters including COD, SCOD, BOD$_5$, TKN, NH$_4^+$-N, TP and TSS were sampled and investigated.

COD removal. The average SCOD values in the settled sewage and the four compartments were 109, 24, 14, 18 and 11 mg l^{-1}, respectively. Most of the SCOD was removed in the anoxic compartment due to denitrification, while SCOD removal in the second and third aerobic compartments was low. The average SCOD removal efficiency was 90.0%.

Nitrification. The average diurnal NH$_4^+$-N concentrations in the settled sewage and four compartments were 25.6, 10.3, 2.0, 1.4 and 1.6 mg NH$_4^+$-N l^{-1}, respectively, indicating that most of the NH$_4^+$-N in the feed was removed in the first aerobic compartment, and the contributions of the second and third compartments in NH$_4^+$-N oxidation were marginal. The average NH$_4^+$-N removal efficiency was 93.8%. The NH$_4^+$-N concentration profiles in the three aerobic compartments were « 5 mg NH$_4^+$-N l^{-1}, the feed water quality requirement of the NEWater plants. The specific nitrification activity correlated well with the ambient NH$_4^+$-N mass load.

Denitrification. The average NO$_3^-$-N concentrations in the four compartments were 1.2, 6.9, 8.2 and 8.9 mg NO$_3^-$-N l^{-1}, respectively. The average NH$_4^+$-N-based denitrification efficiency was 62.1%. The anoxic compartment contributed 34.0%, the final clarifier sludge blanket contributed 23.0% and the aerobic compartments contributed 5.1%.

Mass balance

COD. The percentage distribution of influent COD in descending order in the activated sludge process is: 47%, oxygen-COD consumption in the aerobic compartments; 28%, COD assimilation into biomass; 9%, COD removal in the anoxic compartment; 7%, COD removal in the final clarifier; 7%, particulate COD in the final effluent; and 2%, soluble COD in the final effluent. The net sludge yield was 30.2%. The total COD catabolism coefficient was 69.8%. A 26% reduction on carbonaceous oxygen requirement was achieved due to denitrification in the anoxic zone and FC. Taking into account the oxygen requirement for autotrophic ammonia–nitrogen oxidation, a 12% total oxygen demand reduction was achieved.

Nitrogen. The percentage distribution of influent nitrogen in descending order in the activated sludge process is: 26%, NO_3^--N removal in the anoxic compartment; 24%, NO_3^--N removal in the final effluent; 20%, NO_3^--N removal in the final clarifier; 18%, nitrogen assimilation into sludge cells; 5%, NH_4^+-N in the final effluent; 4%, suspended TKN in the final effluent; and 3%, soluble organic nitrogen in the final effluent.

This detailed information and investigation on the diurnal feed, carbonaceous and nitrogenous matter conversion in individual parts of as well as the whole full-scale activated sludge process exhibited a clear dynamic picture of the system performance, and helped in giving an insightful understanding of the process. These data were used in the design and study of the laboratory-scale process and the parameter verification of modeling.

REFERENCES

Burrows L. J. , West J. R., Forster C. F. and Martin A. (2001) Mixing Studies in An Orbal Activated Sludge System, Wat. SA, **27**(1), 79-83.

Concha L. and Henze M. (1992) Advanced Design and Operation of Municipal Wastewater Treatment Plant. Technologies for Environmental Protection Report 1, EUR 16869 EN.

Concha L. and Henze M. (1996) Advanced Design and Operation of Municipal Wastewater Treatment Plant. Technologies for Environmental Protection Report 10, EUR 16869 EN.

Cao Y.S., Ang C.M. and Zhao W. (2003) Performance Analysis of Full-and Laboratory-Scale Activated Sludge Process at Kranji and Bedok Water Reclamation Plants. Technical Report, Ref no. SUI/2001/030/TRTP1.

Ekama G., Barnard J., Gunthert F., Kreb P., McCorquodale J.A., Parkers D.S. and Waihlberg E. (1997) Secondary Setting Tanks: Theory, Modeling and Operation, IAWQ Report No. 6, IAWQ, London.

Gragy C.P.L., Daigger G. T. and Lim H.C. (1999) Biological Wastewater Treatment, 2nd ed., Marcel Dekker. New York.

Haarbo A., Harremoës P. and Thirsing C (2001) Kinetic Start-Up Performance of Two Large Treatment Plants for Nutrient Removal. Wat. Sci. Tech. **43** (11), 153–160.

Henze M., Dupont R., Grau P. and de la Sota (1993) Rising Sludge in Secondary Settlers due to Denitrification, Wat. Res. 27(2), 231-236.

Henze M., Harremoës P., Janseen J. and Arvin E. (1997) Wastewater Treatment: Biological and Chemical Process, 2nd ed., Springer, Berlin.

Lessouef A., Payraudeau M., Regalla F. and Kleiber B. (1992) Optimizing Nitrogen Removal Reactor Configurations by On-Site Calibration of the IAWPRC Activated Sludge Model. Wat. Sci. Tech. **25**(6), 105-123.

Meijer S.C.F. (2004) Theoretical and Practical Aspects of Modeling Activated Sludge Processes, PhD Thesis, Delft University of Technology, The Netherlands.

Water Environment Federation (WEF) (1992) Design of Municipal Waste Water Treatment Plant, Manual of Practice, No. 8.

Metcalf and Eddy Inc. (2003) Wastewater Engineering: Treatment, Disposal and Reuse, 4th ed., McGraw-Hill, Washington, USA.

Siegrist, H and Tschui, M. (1992) Interpretation of Experimental Data with Activated Sludge Model No. 1 and Calibration of the Model for Municipal Wastewater Treatment Plants. Wat. Sci. Tech. **25**(6), 167-183.

van Veldhuizen H.M., van Loosdrecht M.C.M and. Heijnen J.J (1999) Modeling Biological Phosphorus and Nitrogen Removal in a Full Scale Activated Sludge Process. Wat. Res. **33**(16), 3459-3468.

4

Scaled-down laboratory experimentation

One objective of this study is to explore the feasibility of using laboratory experiment to investigate the performance and optimization of full-scale activated sludge process, and of modeling the full-scale activated sludge process by using the data obtained from laboratory experiment. Thus, a significant number of laboratory experiments were conducted to:

i   simulate the performance of the full-scale activated sludge processes of Bedok and Kranji WRPs and compare the results of the laboratory-scale processes with those of the full-scale processes (Cao *et al.*, 2003; Cao *et al.*, 2004);

ii   study other BNR activated sludge processes such as Bardenpho, Step-Feed etc., (Cao *et al.*, 2004); and

iii   optimize the BNR activated sludge process in a warm climate (Cao *et al.*, 2004).

Section 4.1 firstly defines the scale-down principle developed for activated sludge process development in this study, followed by discussion of the experimental design and set-up (Section 4.2.1). Sections 4.2.2 to 4.2.8 present the results of the experiments to simulate the MLE activated sludge process of Phase IV of Bedok WRP, which was described in Chapter 3. The laboratory-scale study on the modified design of the BNR activated sludge process, with an interest in the optimization of the appropriate aerobic SRT and the enhancement of denitrification in warm climates, based on the feed conditions of Phase III of Kranji WRP (Section 2.3.1), are presented in Section 4.3. The results with the feed conditions of Battery B of Seletar WRP are presented in Section 5.4.1.

4.1 SCALED-DOWN LABORATORY ACTIVATED SLUDGE SYSTEM

4.1.1 The scale-down principle in the activated sludge process

Laboratory experimentation is a cost-effective way adopted for the investigation of the improvement and optimization of an existing process and the development of new processes. Considerable efforts have been made to ensure that the outcomes of the laboratory experiment will be able to simulate the performance of the full-scale system under investigation. Earlier, the popular approach was to design the laboratory-scale system based on the principle of similarity of geometric dimension ratios, and then later based on dimensional analysis (Johnstone and Thring, 1957).

In the mid 1980s, the mechanistic scale-down principle was developed in process scale-up, and it has been successfully applied in the chemical and food industries. The Kluyver Institute for Technology, Delft University of Technology, made important contributions in this area (Oosterhuis, 1984; Kossen and Oosterhuis, 1985; Sweere, 1987; Luyben, 1993; Cui, 1997). During the scale-up process, the intrinsic kinetics remain unchanged while the process kinetics comprising intrinsic kinetics and mass transfer may change due to the enlarged scale. The key is to identify the step(s) of the mass transfer, which change(s) in the scale-up process, and to investigate the impact of these changes on the performance of full-scale process. 'Regime analysis', which is carried out through time constant comparison aiming at determining the 'ruling regime' that governs the system performance, was developed and has become a useful tool in scale-up and down. Under this method, essentially, a laboratory experiment designed to study full-scale process should be operated under the same 'ruling regime' as that of the full-scale system; otherwise, it will be necessary to understand the impact of regime change to process development. Therefore, the identification of the key parameters and conditions, which determine the full-scale system performance, is a pre-condition for successful scale-down experiment. This requires an early participation of process engineers in laboratory research as an effective way to shorten the process development period (Goldstein et al., 1991).

The development in mechanistic scale-down in the water sector is slow. The main approach is still geometric ratio similarity (Schmidtke and Smith, 1987); and in most cases, this approach requires several stages of scale-up with pilot-scale plants of different sizes to test scale-up effects. It is costly and time consuming. Grady (1993) described the principles and methodologies of scale-down and scale-up in wastewater treatment process. He emphasized the roles of reactor engineering, bench- and pilot-scale tests and models in scale-up. However, no general guidelines were given.

In this study, three categories of parameters and conditions which govern the performance of an activated sludge process are being defined as ruling regimes of scale-down, as follows:

i   **Feed conditions**. These include diurnal flow pattern and compositions. The wastewater used in the scale–down study should be from the same source as the full-scale process. Compositions refers to both the conventional parameters and COD fractions (finger-prints);

ii   **Bioreactor system**. The critical factor is the hydraulic flow pattern in the reactor. A normal way is to keep similar configuration including the ratios of the length, width and depth between the full- and laboratory-scale systems. In the past ten years, Residence Time Distribution (RTD) has been applied to investigate the non-ideal flow of the full-scale activated sludge system (Siegrist and Tschui, 1992; Burrows et al. 1999; Burrows et al. 2001; Kjellstrand et al. 2005). Chemical reaction engineering knowledge helps in this respect. A laboratory-scale system can be designed to incorporate non-ideal flow in the actual process. For example, three Continuously Stirred Tank Reactors (CSTRs) in series in the laboratory-scale activated sludge system were able to approximately simulate a plug flow but with certain longitudinal mixing (Levenspiel, 1972); and

iii **Biochemical environment.** The most essential parameter is the SRT, which plays a critical role in the selection of microorganisms to be either retained in or washed out of the reactor. The volumetric ratios of aerobic, anoxic and anaerobic compartments, MLVSS concentration, and temperature, DO and pH etc., in the reactor are also important parameters. Other factors include HRT, which determines the duration of the substrate-biomass contact and operational parameters such as internal mixed liquor recycle and return sludge recycling ratios etc.

Based on the scale-down principle, a laboratory experiment should be designed with similar ruling regimes defined by the three categories of parameters as those of the full-scale process under investigation. Thus, it is expected that the performance of the laboratory experiment enables the prediction of the performance of the full-scale activated sludge process.

4.2 LABORATORY SIMULATION OF THE ACTIVATED SLUDGE PROCESS OF BEDOK WRP

4.2.1 Experimental design, set-up and feed

The laboratory experiment was designed to simulate the MLE activated sludge process of Bedok WRP.

4.2.1.1 Feed conditions

The typical diurnal hydraulic conditions in Bedok WRP introduced in Section 2.3.1 were adopted in this experimental study. Both the settled sewage and sludge were taken from the site. The settled sewage was stored in a refrigerator at a temperature below 4 $^{\circ}$C to ensure that the feed sewage composition remained unchanged before being pumped into the laboratory-scale reactor system. Since the SCOD of the sampled sewage was lower than the typical sewage, acetate was added to regulate the SCOD level. Separate COD and ammonia-nitrogen concentrations of the sewage feed were used during the hydraulic flow peak period and the non-peak period as shown in Table 4.1.

The diurnal flow, and the COD and ammonia-nitrogen mass loading rates were pumped by a programmed peristaltic pump (Model 7550-22, Masterflex, USA) for dynamic feeding. The average flow rate was regulated such that it corresponded to a HRT of 6.2 h. The sole hydraulic peak occurred between 10:00 and 13:00 with a peak factor of 1.4, while low flow occurred between 02:00 and 07:00 with a factor of 0.6. Figures 4.1 and 4.2 show the diurnal flow, COD and ammonia-nitrogen mass loading rates.

Table 4.1. Conventional parameters of the raw settled sewage feed and major feed characteristics during the peak concentration period and the non-peak concentration period.

Parameters (mg l^{-1})	COD	SCOD	TSS	BOD$_5$	TKN	NH$_4^+$-N	TP	ALK
Settled sewage	280	125	99.7	49	47.0	41.3	19.1	206.4
Peak	280	146	84.7	82	40.0	35.1	16.2	193.3
Non-Peak	230	125	67.8	73	32.0	28.1	13.0	168.8

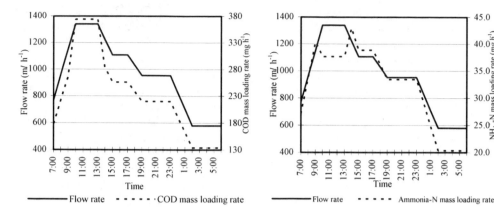

Figure 4.1. Diurnal influent flow and COD mass loading rate profiles of the laboratory-scale MLE activated sludge process of Phase IV of Bedok WRP.

Figure 4.2. Diurnal influent flow and NH_4^+-N mass loading rate profiles of the laboratory-scale MLE activated sludge process of Phase IV of Bedok WRP.

4.2.1.2 Bioreactor system

Considering the non-ideal flow caused by longitudinal mixing and dead space in the full-scale activated sludge process (Burrow *et al.*, 2001), a system comprising three CSTRs in series, which is the minimum number required for simulation of an approximate plug-flow pattern (Levenspiel, 1972), was adopted in the laboratory-scale reactor system to simulate the hydraulic flow pattern on site. As shown in Figure 4.3, the system consists of one anoxic reactor and two aerobic reactors in series, where each reactor is a single CSTR. The effective volume of the anoxic reactor was 1.41 l and the volume of each aerobic reactor was 2.12 l, so the total effective volume of the three reactors was about 5.65 l. The anoxic volumetric ratio was 25% of total volume, which was similar to the site condition, and the aerobic volumetric ratio of each aerobic reactor was 37.5% of the total volume. Both the RAS and MLR ratios were 100% of the average influent flow rate.

4.2.1.3 Biochemical environment

The designed total SRT was 10 d, which was similar to the site data. It was manipulated by hydraulic control through removal of a known volume of mixed liquor from each reactor on a daily basis. Based on a total SRT of 10 d, neglecting MLVSS concentration difference in each reactor, and taking into account the volumes of the anoxic and aerobic reactors, the SRT_{ANO} and SRT_{AER} were 2.5 d and 7.5 d, respectively. The contents in the three reactors were maintained at a temperature of $30 \pm 1\ ^0C$. Stirring, at 120 rpm, was also applied in the anoxic reactor for the purpose of mixing. Air was supplied by an air pump (Beetle 12000), and the DO concentration in each of the two aerobic reactors was maintained in a range between 1.0 and 2.0 mg $O_2\ l^{-1}$.

For most of the laboratory experiments, the duration of each experiment operated under certain conditions was at least three SRTs to ensure the system reached a 'well-developed' or 'stable state' even under dynamic feed conditions. Samples taken for analysis were then representative of the performance of the process. Sampling and analysis were carried out according to the methods and procedures as described in the previous chapters.

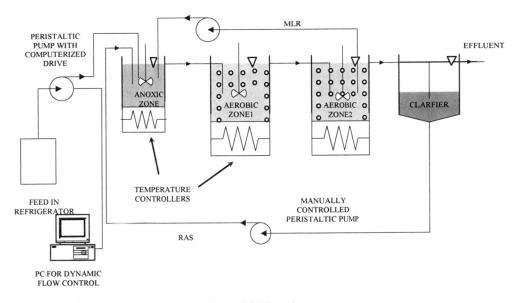

Figure 4.3. Laboratory configuration of the CSTR series system.

Figure 4.4 shows a laboratory-scale activated sludge system consisting of a program controlled feed system, three CSTRs in series, with an effective volume of about 5.6 l only, and a final clarifier adopted in this study. The intention was to explore the feasibility of using this system at such scale to simulate directly the performance of a single full-scale activated sludge unit with an effective volume of 8,724 m^3. The magnitude of the difference in scale was more than 1,000,000 times, which has never before been reported in literature related with wastewater process development.

Figure 4.4. Laboratory-scale activated sludge system used to simulate the performance of the full-scale activated sludge process of Bedok WRP.

4.2.2 Carbonaceous matter removal

COD removal was studied mainly through SCOD removal. As shown in Figure 4.5, the peaks in the anoxic reactors occurred between 09:00 and 12:00 and were dampened significantly in the first and second aerobic reactors, reflecting the effects of SCOD (and COD) mass load shock. The average SCOD values in the feed and three reactors were 128, 15, 12 and 12 mg l^{-1}, respectively. Taking the average SCOD value at the inlet of the anoxic reactor as 40 mg l^{-1}, it can be concluded that most of the SCOD was removed in the anoxic reactor through denitrification and some of the remaining SCOD through carbon oxidation in the first aerobic reactor. The contribution of the second aerobic reactor in terms of SCOD removal was minimal. This profile was similar to the case in the full-scale process (Section 3.3.2). Given the average SCOD value of 128 mg l^{-1} in the feed and the average values of the effluents of the reactors, the SCOD removal efficiency was 90.6 %, which was calculated by Equation 3.3 and was almost similar to that (90.0 %) of the full-scale activated sludge process of Bedok WRP (Section 3.3.3).

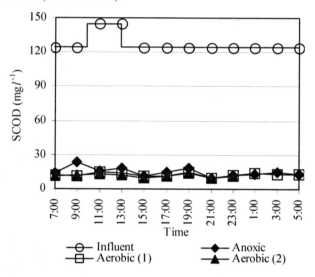

Figure 4.5. SCOD concentration profiles of the influent and the three reactors of the laboratory-scale MLE activated sludge process of Phase IV of Bedok WRP.

4.2.3 Nitrification

Figure 4.6 shows that one peak occurred in the anoxic reactor from the morning until early afternoon corresponding to both the hydraulic flow peak and ammonia-nitrogen mass loading rate peak in the feed. Similar to the full-scale process (Figure 3.7), the peak then dampened in the first aerobic reactor and almost disappeared in the second aerobic reactor.

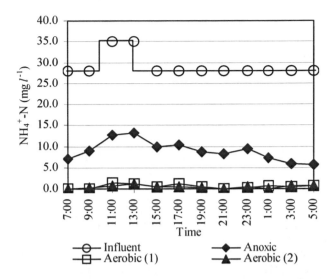

Figure 4.6. NH_4^+-N concentration profiles of the influent and the three reactors of the laboratory-scale MLE activated sludge process of Phase IV of Bedok WRP.

The average diurnal ammonia-nitrogen concentrations in the effluents of the feed and three reactors calculated were 29.3, 8.9, 0.6 and 0.5 mg NH_4^+-N l^{-1}. The overall nitrification efficiency based on NH_4^+-N, which was calculated by Equation 3.4, was 98.3 %. Compared with the average values of 8.9, 2.0, 1.4 and 1.6 mg NH_4^+-N l^{-1} in the four compartments of the full-scale system of Phase IV of Bedok WRP (Section 3.3.4) the similarity was obvious, i.e. the nitrification occurred mainly in the first aerobic reactor although the NH_4^+-N concentrations in the aerobic reactors of the laboratory experiment were slightly lower than those of the full-scale system due to the oxygen supply conditions. With regards to SCOD and ammonia removal, the functions of the final aerobic reactor were limited, implying, probably, that the SRT of 7.5 d for the aerobic zone might be more than sufficient for nitrification in the warm conditions of Singapore.

The specific nitrification rates shown in Figures 4.7(a) and (b), which were obtained through dividing the NH_4^+-N mass oxidation rate (calculated by Equation 3.1) by (MLVSS x V_{AER}), indicated that the rate in the first aerobic reactor varied between 3.2 and 10.0 mg NH_4^+-N (g VSS)$^{-1}$h^{-1}, which was about 20% less than that of the first aerobic compartment of the full-scale process of Phase IV of Bedok WRP (Section 3.3.4). The low F/M ratio of the laboratory-scale system, due to the higher volumetric ratio of the aerobic reactor (37.5%) in the laboratory-scale system compared to that of the full-scale system (25%), was the main cause. However, the rate in the second aerobic reactor was close to that of the last compartment of the full-scale system.

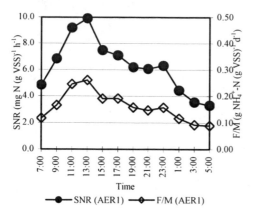
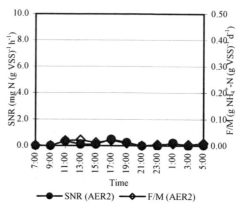

Figure 4.7(a). Variations of the specific nitrification rate with the NH_4^+-N-based F/M ratio in the first aerobic reactor of the laboratory-scale MLE activated sludge process of Phase IV of Bedok WRP.

Figure 4.7(b). Variations of the specific nitrification rate with the NH_4^+-N-based F/M ratio in the second aerobic reactor of the laboratory-scale MLE activated sludge process of Phase IV of Bedok WRP.

4.2.4 Denitrification

4.2.4.1 Denitrification in the activated sludge tanks

As shown in Figure 4.8, the NO_3^--N concentration in the anoxic reactor was almost non-detectable from 07:00 to 02:00 and then increased to about 1.0 mg NO_3^--N l^{-1} during the base flow period when S_S and X_{S1} in the sewage were low (Section 2.3.3.2). The diurnal average concentration was 0.3 mg NO_3^--N l^{-1} only. It was lower than 1.2 mg NO_3^--N l^{-1}, the corresponding diurnal average concentration on site (Section 3.3.5.1). Acetate addition to the laboratory sewage feed, which increased the readily biodegradable COD (S_S) fraction, could be the reason.

The diurnal average concentrations in the first and second aerobic reactors were 8.2 and 9.2 mg NO_3^--N l^{-1}, which were comparable to the range between 6.9 and 8.9 mg NO_3^--N l^{-1} in the full-scale system (Section 3.3.5.1). The nitrate-nitrogen concentration increase in the second aerobic reactor was only 1.0 mg NO_3^--N l^{-1} on a daily basis, corresponding to modest ammonia reduction, which was similar to that of the activated sludge process of Phase IV of Bedok WRP. Thus, significant increases in nitrate-nitrogen concentrations were observed as a result of ammonia oxidation in the first aerobic compartment.

Figure 4.9 shows the NO_3^--N mass removal rate and influent-based NO_3^--N concentration removal profiles, which was calculated by 3.1 and 3.2, in the anoxic reactor. The mass removal rate reached the maximum of 17.0 mg NO_3^--N h^{-1} at 17:00 and 23:00 at night while the maximum influent-based nitrate-nitrogen removal concentration was 23.0 mg NO_3^--N l^{-1}. Both were higher than those of the site data. The specific denitrification rate, which was obtained through dividing the NO_3^--N mass removal rate (calculated by Equation 3.1) by (MLVSS x V_{ANO}) reached the maximum of 5.7 mg NO_3^--N (g VSS)$^{-1}$ h^{-1} at about 17:00 as shown in Figure 4.10. This rate was higher than 4.2 mg NO_3^--N (g VSS)$^{-1}$ h^{-1}, the maximum rate in the first anoxic compartment of the full-scale process of Phase IV of Bedok WRP (Section 3.3.5.1). Acetate addition to the sewage feed of the laboratory experiment could be the main reason.

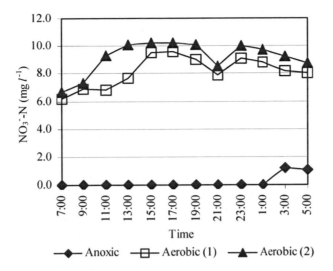

Figure 4.8. NO₃-N concentration profiles of the three reactors of the laboratory-scale MLE activated sludge process of Phase IV of Bedok WRP.

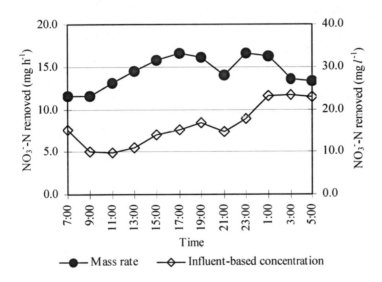

Figure 4.9. NO₃-N removed, in terms of the mass removal rate and influent-based concentration, in the anoxic reactor of the laboratory-scale MLE activated sludge process of Phase IV of Bedok WRP.

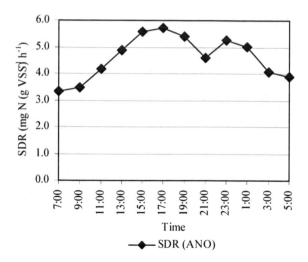

Figure 4.10. Specific denitrification rates in the anoxic reactor of the laboratory-scale MLE activated sludge process of Phase IV of Bedok WRP.

4.2.4.2 Denitrification in the final clarifier

As shown in Figure 4.11, the nitrate-nitrogen concentration in the RAS was the lowest, indicating that denitrification occurred in the final clarifier. The NO_3^--N removal rate calculated by Equation 3.1 reached a maximum of 7.2 mg NO_3^--N h^{-1} at about 11:00 (Figure 4.12) when the NO_3^--N concentration at the clarifier inlet nearly reached the peak. This value was about 3 mg NO_3^--N h^{-1} less compared to the full-scale data, probably due to the thinner level of sludge blanket in the laboratory experiment. Given the average NO_3-N concentration of the final effluent was 8.4 mg NO_3-N l^{-1} (Section 4.2.6), the average overall denitrification efficiency based on NH_4^+-N, calculated by E3.6, was 70.8%.

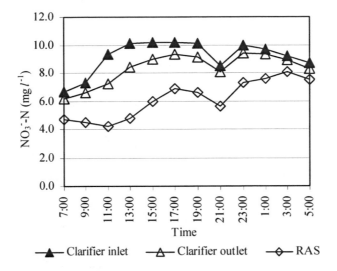

Figure 4.11. NO_3^--N concentration profiles at the inlet and outlet of the clarifier and of the RAS of the laboratory-scale MLE activated sludge process of Phase IV of Bedok WRP.

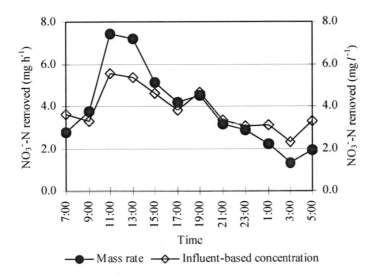

Figure 4.12. NO_3^--N removed, in terms of the mass removal rate and influent-based concentration, in the clarifier of the laboratory-scale MLE activated sludge process of Phase IV of Bedok WRP.

4.2.5 pH and alkalinity

Figure 4.13 shows that the pH reduced significantly, as a result of nitrification, from 7.22, the average in the anoxic reactor, to 6.87 and 6.88, the respective average values in the first and second aerobic reactors. The difference in pH in the two aerobic reactors was insignificant as nitrification in the second reactor was marginal. The average pH of 6.88 in the second aerobic rector was higher than 6.25, the average pH in the last compartment of the full-scale activated sludge process (Section 3.3.6), most likely due to more nitrate-nitrogen reduction as a result of acetate addition. As shown in Figure 4.14, the alkalinity profiles were consistent with the pH profiles, and the diurnal average alkalinity in the second aerobic reactor was 32.7 mg as $CaCO_3$ l^{-1}.

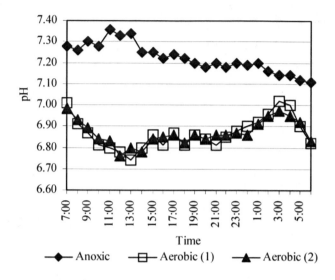

Figure 4.13. pH profiles of the three reactors of the laboratory-scale MLE activated sludge process of Phase IV of Bedok WRP.

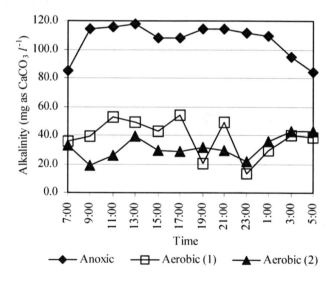

Figure 4.14. Alkalinity profiles of the three reactors of the laboratory-scale MLE activated sludge process of Phase IV of Bedok WRP.

4.2.6 Effluent quality

Figure 4.15 shows the nitrogenous component concentrations in the final effluent collected during various time intervals. The diurnal weighted average ammonia-nitrogen, nitrate-nitrogen and inorganic ammonia-nitrogen (soluble TKN – NH_4^+-N) concentrations were 0.6, 8.4 and 0.6 mg N l^{-1} respectively, and were comparable to those of the full-scale activated sludge process of Bedok WRP (Section 3.3.7).

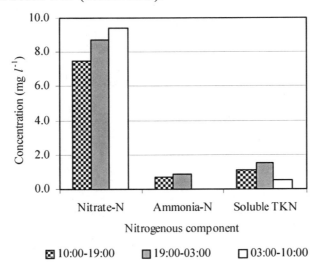

Figure 4.15. Nitrogenous components concentrations of the final effluent of the laboratory-scale MLE activated sludge process of Phase IV of Bedok WRP during various time intervals.

4.2.7 Mass balance and yield coefficients

4.2.7.1 COD balance

Approaches used in the COD mass balance calculation of the full-scale activated sludge process of Phase IV of Bedok WRP (Section 3.3.8.1) were similarly adopted here. Table 4.2 summarizes the results calculated on a 24-h basis.

Table 4.2. Distribution of the COD (mg COD d^{-1}) in the laboratory scale MLE activated sludge process of Phase IV of Bedok WRP.

Denitrified COD in the anoxic reactor	Denitrified COD in the final clarifier	Aeration Removed COD (RO$_H$)	Anabolized COD$_{WAS}$	COD in final effluent		Influent COD
				Soluble	Particulate	
994	266	2 061	1 619	160	410	5 510

* Calculated according to a SRT of 10 d and 10% COD content of the VSS.

Figure 4.16 shows the percentage distribution based on the values of Table 4.2: 18%, COD removal in the anoxic reactor; 30%, COD assimilation into biomass; 37%, oxygen-COD consumption in the aerobic reactors; 5%, COD removal in the final clarifier; 7%, particulate COD in the final effluent; and 3%, soluble COD in the final effluent. Compared with the full-scale data in Section 3.3.8.1, COD assimilation and particulate COD in the final effluent were correspondingly similar. However, the percentage of COD removed in the anoxic reactor in the laboratory experiment was higher than that of the full-scale system, which led to a lower oxygen demand in the aerobic reactor of the laboratory. The addition of acetate to the sewage feed of the laboratory experiment could be the main cause.

The net sludge yield calculated by Equation 3.7 was 32.8% and the catabolized yield calculated by Equation 3.8 was 67.2%. Both were close to the full-scale data of 30.2% and 69.8%, respectively.

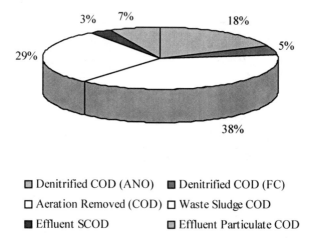

□ Denitrified COD (ANO)   ■ Denitrified COD (FC)
□ Aeration Removed (COD)   □ Waste Sludge COD
■ Effluent SCOD   ▨ Effluent Particulate COD

Figure 4.16. Percentage distribution of the influent COD in the MLE activated sludge process of Phase IV of Bedok WRP.

4.2.7.2 Nitrogen balance

As indicated in Table 4.3, the total nitrogen input to the system of 775 mg N d^{-1} was the sum of the nitrogen removed in the anoxic reactor and final clarifier, that assimilated into cells (wasting), and that discharged in the effluent. The calculated total nitrogen input was only 1.0% higher than the influent TKN load to the system, illustrating the reliability of the measurement.

Table 4.3. Distribution of the nitrogen (mg N d^{-1}) in the laboratory-scale MLE activated sludge process of Phase IV of Bedok WRP.

Nitrogen removed in anoxic reactor	Nitrogen removed in final clarifier	Nitrogen in wasted sludge (WAS)*	Nitrogen in final effluent				Calculated influent TN
			NH_4^+-N	NO_3^--N	Sol. Org-N	Susp. TKN**	
346	93	135	11	154	9	27	775

* Calculated according to a SRT of 10 d and 10% nitrogen content of the VSS
** Calculated from the TSS in the effluent and 10% nitrogen content of the VSS

Figure 4.17 shows the percentage distribution of nitrogen in the laboratory-scale activated sludge process. The percentage distribution in descending order is: 45%, NO_3^--N removal in the anoxic reactor; 20%, NO_3^--N in the final effluent; 17%, nitrogen assimilation into sludge cells;

12%, NO_3^--N removal in the final clarifier; 4%, suspended TKN in the final effluent; 1%, NH_4^+-N in the final effluent; and 1%, soluble organic nitrogen in the final effluent.

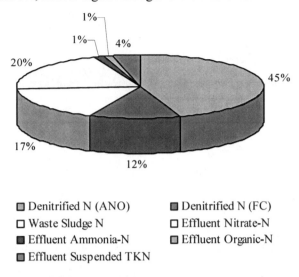

Figure 4.17. Percentage distribution of the influent nitrogen in the laboratory-scale MLE activated sludge process of Phase IV of Bedok WRP.

The nitrification efficiency based on the nitrogen balance where nitrogen assimilation was taken into account calculated by Equation 3.9 was 92.7%. Similarly, the overall denitrification efficiency based on nitrogen balance, taking nitrogen assimilation into account calculated by Equation 3.10 was 74.0%, of which, 58.3% was attributed to the anoxic reactor and 15.7% to the final clarifier. Table 4.4 records the nitrification and denitrification efficiencies calculated based on ammonia-nitrogen and total nitrogen.

Table 4.4. Nitrification and denitrification efficiencies (%) of the laboratory-scale MLE activated sludge process of Phase IV of Bedok WRP.

Nitrification		Denitrification in the anoxic reactor and clarifier (based on total nitrogen)		Total denitrification	
Based on NH_4^+-N	Based on total nitrogen	Anoxic zone	Final clarifier	Based on NH_4^+-N	Based on total nitrogen
98.3	92.7	58.3	15.7	70.8	74.0

4.2.8 Performance comparison between the full- and laboratory-scale systems

Tables 4.5, 4.6 and 4.7 compile the key performance indicator parameters of both the full- and laboratory-scale processes showing the comparison between laboratory-scale simulation and the full-scale activated sludge process performance. As indicated in Table 4.5, the diurnal average concentration profiles of SCOD, NH_4^+-N and NO_3^--N in the anoxic and aerobic compartments were quite similar: SCOD and nitrate-nitrogen reduction occurred mainly in the anoxic compartment while nitrification occurred in the first aerobic compartment. The SCOD reduction and nitrification in the last two aerobic compartments of the full-scale activated sludge process of Phase IV of Bedok WRP and in the second aerobic reactor of the laboratory-scale system were marginal. Tables 4.5 and 4.6 show that the ammonia nitrogen concentration in the second aerobic reactor of the laboratory-scale system was only slightly lower (< 1.5 mg N l^{-1}) than that of the full-scale system, as the oxygen supply in the laboratory-scale system was sufficiently more than that in the full-scale activated sludge process of Phase IV of Bedok WRP.

Table 4.5. Key diurnal average concentrations in the full- and laboratory-scale processes (mg l^{-1}).

Parameter	System	ANO	AER (1)	AER (2)	AER (3)
SCOD	Full-Scale	24	14	18	11
	Laboratory	15	12	12	
NH_4^+-N	Full-Scale	8.9	2.0	1.4	1.6
	Laboratory	8.9	0.6	0.5	
NO_3^--N	Full-Scale	1.2	6.9	8.2	8.9
	Laboratory	0.3	8.2	9.2	
Alkalinity (as $CaCO_3$)	Full-Scale	92.6	47.9	41.0	37.0
	Laboratory	106.2	39.7	32.7	

Table 4.6. SCOD, NH_4^+-N and nitrogen removal efficiencies of the laboratory- and full-scale systems (in %).

	SCOD	Nitrification		Denitrification		Remarks
		NH_4^+-N based	TN based	NH_4^+-N based	TN based	
Laboratory-scale test	90.6	98.3	92.7	70.8	74.0	The DO in the activated sludge process of Bedok WRP was lower and affected nitrification
Full-scale test	90.0	93.9	85.0	62.1	65.6	

Table 4.7. COD and nitrogen mass balance in the full- and laboratory-scale systems (%).

Parameter	System	ANO	AER	Assimilation	Effluent		Secondary clarifier
					Soluble	Particulate	
COD	Full-Scale	9	47	28	2	7	7
	Laboratory	18	30	37	3	7	5
Nitrogen	Full-Scale	26	NA	18	32	4	20
	Laboratory	45	NA	17	22	4	12

Table 4.7 indicated that the distribution of the COD utilized for assimilation, in the final effluent and that removed in the secondary clarifier in the two processes of different scales, were close to each other. The COD removal in the anoxic reactor at the laboratory-scale process was higher than that of the full-scale system; and acetate addition to the sewage feed of the laboratory-scale study could be the main cause. Similarly, for nitrogen, the

percentages of biomass assimilation in the two systems were close to each other. The major difference here was that the NO_3^--N denitrified in the anoxic reactor of the laboratory-scale system was higher than that of the full-scale system due to the high denitrification efficiency in the laboratory experiment, which was consistent with the high COD removal percentage in the anoxic reactor of the laboratory-scale process. This illustrated that COD fractions (finger prints) of the feed should also be included in addition to the conventional parameters.

The comparisons of both system performance in kinetics (reaction rates) and mass flow and balance illustrated that the laboratory-scale activated sludge system designed based on scale-down principles was able to simulate the performance of the full-scale process.

4.3 OPTIMIZATION OF THE BNR ACTIVATED SLUDGE PROCESS IN WARM CLIMATES

Laboratory experiments were performed focusing on an appropriate SRT for nitrification, denitrification, and comparison of different BNR processes such as Bardenpho, Step-Feed and MLE, etc., for the optimization of the BNR activated sludge process in warm climates (Cao *et al.* 2004). The results of the studies on the appropriate SRT of MLE process with the feed conditions of Phase IV of Bedok WRP are presented and discussed in the following sections.

4.3.1 Conceptual design, experimental set-up and feed

4.3.1.1 Warm temperature and shorter aerobic SRT

Elevated year-round water temperatures ($30 \pm 1°C$) and relatively limited readily biodegradable COD content are two typical characteristics of the sewage in Singapore. Temperature has a direct effect on the SRT required for carbonaceous matter removal and, especially, ammonia oxidation. Bacterial growth rates increase with temperature (Stainer *et al.*, 1986) and generally speaking, the higher the temperature, the shorter the SRT required for biological processes (Rittmann and McCarty, 2002). The aerobic SRT required for nitrification is temperature dependent, as shown in Figure 4.18 (Henze *et al.*, 1997). Nevertheless, little practical information is available on the aerobic SRT required for nitrification under warm conditions since most information is for temperate or cold climatic conditions (EPA, 1993). The aerobic SRTs adopted in the existing activated sludge processes of water reclamation plants in Singapore are in the range of 7 to 8 days, which is typically used for the BNR systems under temperate conditions (EPA, 1993). Thus, the question arising is: can the aerobic SRT be reduced in the presence of high temperatures such as those of Singapore?

Applying a shorter aerobic SRT means a smaller aeration tank volume and less aeration energy consumption that can account for up to 50% of the electrical supply in wastewater treatment (Metcalf & Eddy Inc., 2003). The results of the full- and laboratory-scale process studies presented so far in this book have demonstrated that most of the soluble COD and NH_4^+-N were removed in the initial half of the aerobic tanks, thus throwing some light on the question and indicating a possibility for further reduction of the aerobic SRT.

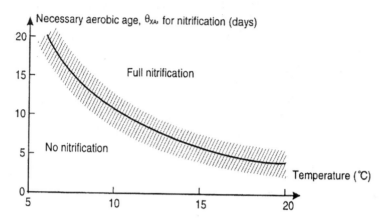

Figure 4.18. Relationship between temperature and necessary aerobic SRT to achieve nitrification in activated sludge plants (adopted from Henze *et al.*, 1997. Copyright © Springer with permission).

The amount and biodegradability of the COD in the sewage impose large effects on the denitrification efficiency of the BNR process. Sewage characterization (Section 2.3.3) showed that the easily biodegradable COD (S_S) and rapidly hydrolysable biodegradable COD (X_{S1}) of Bedok and Kranji WRPs were 'diluted' or 'very diluted' compared to the reported range for several European countries, even though the total COD and ammonia-nitrogen concentrations were within the 'moderate' range. The limited readily biodegradable COD could lead to low denitrification efficiency, which is unfavorable to the stability of the system as the alkalinity of the settled sewage in Singapore is low. Therefore, an appropriate aerobic SRT and effective utilization of slowly biodegradable and intracellular COD as electron donors to enhance denitrification become the main topics in process design and optimization.

4.3.1.2 Experimental design

The aerobic SRT chosen in the laboratory-scale optimization study was 3.75 d, which was only half of that of the existing full-scale systems. To promote the utilization of slowly biodegradable COD, the anoxic volumetric ratio was increased from the existing ratio of 20-25% to 62.5% of the total volume of the activated sludge reactor. In practice, this increase in anoxic volume could be obtained by the reduction of the aerobic volume.

Configuration of MLE process. Figure 4.19 shows the schematic diagram of the MLE configuration and key operation parameters. The HRT was 10 h and total SRT was 10 d. The anoxic volumetric ratio was 62.5% consisting of two anoxic reactors in series and corresponded to an anoxic SRT of 6.25 d. The aerobic volumetric ratio was 37.5%, corresponding to an aerobic SRT of 3.75 d, and only half of that of the current full-scale ratio and value, respectively. Both the RAS and MLR ratios were 100% of the average influent flow rate.

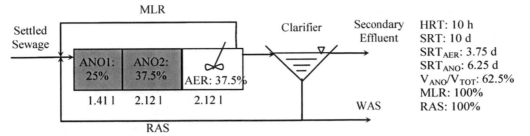

Figure 4.19. Schematic diagram of the laboratory-scale MLE activated sludge process in the optimization study.

Feed. Feed characterization data from the Kranji WRP (Section 2.3.1) were used in the optimization study. Sludge and settled sewage were taken from Kranji WRP. Figures 4.20(a) and 4.20(b) show the diurnal hydraulic flow and typical COD and NH_4^+-N mass loading rates used in the experiment. Both had two peaks, one during the day and the other at night. A low flow was also simulated from late night to early morning. Table 4.8 details the conventional parameters of the settled sewage adopted in the experiment. The NH_4^+-N concentration during the morning peak period was maintained between 34.8 and 36.2 mg NH_4^+-N l^{-1} while for the non-peak period, the concentration was maintained between 29.3 and 29.8 mg NH_4^+-N l^{-1}. As the NO_X^--N concentrations in the influent were negligible, the value of TKN was assumed for TN and the daily nitrogen load was calculated accordingly. The control of the laboratory-scale system and analyses of samples are similar to those in Section 4.2.1.

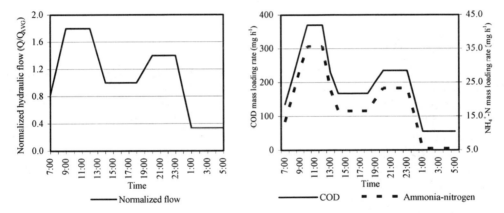

Figure 4.20(a). Normalized hydraulic flow profile of the settled sewage feed used in the laboratory-scale reactors.

Figure 4.20(b). COD and NH_4^+-N mass loading rate profiles of the settled sewage feed used in the laboratory-scale reactors.

Table 4.8. Conventional parameters (mg l^{-1}) of the settled sewage feed during peak and non-peak concentration periods.

Parameter	COD	SCOD	TKN	NH_4^+-N	ALK (as $CaCO_3$)
Peak	376 – 459	140 – 142	47.5 – 49.2	34.8 – 36.2	181.4 – 187.2
Non-Peak	297 – 342	120 – 122	39.2 – 42.7	29.3 – 29.8	155.9 – 158.7

4.3.2 Carbonaceous matter removal

As shown in Figure 4.21, the average values of SCOD in the feed, first and second anoxic and aerobic reactors were 125, 29, 27 and 23 mg COD l^{-1}, respectively. The bulk of the SCOD was removed in the first anoxic reactor, and not much SCOD removal occurred in the second.

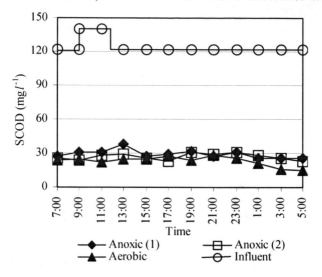

Figure 4.21. SCOD concentration profiles of the influent and the three reactors of the laboratory-scale MLE activated sludge process in the optimization study.

4.3.3 Nitrification

As presented in Figure 4.22, two NH_4^+-N concentration peaks were observed in each of the two anoxic reactors, which reflected the effect of peak NH_4^+-N mass loads. The average NH_4^+-N concentrations of the feed, first and second anoxic and last aerobic reactors were 30.4, 10.5, 10.3 and 0.52 mg NH_4^+-N l^{-1}, respectively. This marginal difference in the NH_4^+-N concentrations of the two anoxic reactors indicated that NH_4^+-N assimilation and hydrolysis of TKN in the second anoxic reactor were balanced. Furthermore, comparing the NH_4^+-N concentrations of the second anoxic reactor with those of the first anoxic reactor, reductions were observed during the morning peak period possibly due to increased NH_4^+-N assimilation for denitrifier growth as a result of elevated COD mass load. At night, as the COD mass load was limited, relatively higher concentrations of NH_4^+-N were released as a result of the hydrolysis of the slowly biodegradable organic matter and decay of microorganisms. The average NH_4^+-N concentration in the aerobic reactor was only 0.52 mg NH_4^+-N l^{-1} and there were no obvious fluctuations of NH_4^+-N concentration recorded during the peak periods. This demonstrated that a complete nitrification was achievable at an aerobic SRT of 3.75 d. Given that the average NH_4-N concentration of the final effluent was 0.84 mg NH_4-N l^{-1} (Section 4.3.6; Table 4.9), the average nitrification efficiency calculated based on NH_4^+-N, which was calculated by Equation 3.4, was 97.4%.

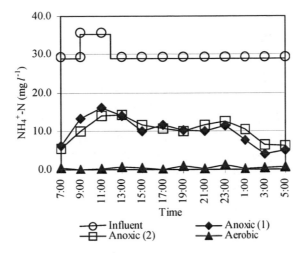

Figure 4.22. NH_4^+-N concentration profiles of the influent and the three reactors of the laboratory-scale MLE activated sludge process in the optimization study.

As shown in Figure 4.23, the specific rate (activity) of NH_4^+-N oxidation in the aerobic reactor, which was obtained by dividing the NH_4-N mass oxidation rate (calculated by Equation 3.1) by (MLVSS x V_{AER}), varied with the NH_4^+-N-based F/M ratio. The specific rate reached a maximum of 7.0 mg NH_4^+-N (g VSS)$^{-1}$h^{-1} at about 11:00 during the morning peak period when the NH_4^+-N-based F/M ratio reached a maximum of 0.17 g NH_4^+-N (g VSS)$^{-1}$d^{-1}, and a minimum of 1.8 mg NH_4^+-N (g VSS)$^{-1}$h^{-1} at about 05:00 during the base flow period when the NH_4^+-N-based F/M ratio reached a minimum of 0.05 g NH_4^+-N (g VSS)$^{-1}$d^{-1}. The magnitude correlation between activities and NH_4^+-N-based F/M ratio was comparable with that of the full-scale data of Phase IV of Bedok WRP (Figure 3.8(a)) and laboratory data simulating the full-scale performance of the activated sludge process of Phase IV of Bedok WRP (Figure 4.7(a)).

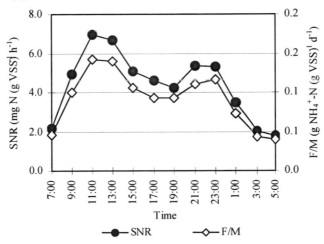

Figure 4.23. Specific nitrification rates and NH_4^+-N-based F/M ratios in the aerobic reactor of the laboratory-scale MLE activated sludge process in the optimization study.

4.3.4 Denitrification

4.3.4.1 Denitrification in the activated sludge tanks

Figure 4.24 shows the NO_3^--N concentration profiles in the three reactors. The average NO_3^--N concentrations in the three reactors were 5.4, 3.5 and 14.6 mg NO_3^--N l^{-1}, respectively. The NO_3^--N concentration reduction of 1.9 mg NO_3^--N l^{-1} between the first and second anoxic reactor demonstrated the enhancement of denitrification in the second anoxic reactor, which occurred mainly between 13:00 and 22:00 and between 01:00 and 05:00. The flows were low during these two periods, and possibly, led to an increase in the MLSS concentration and a longer retention time, which favored the utilization of the particulate COD. The average NO_3^--N concentration increased from 3.5 mg NO_3^--N l^{-1} in the second anoxic reactor to 14.6 mg NO_3^--N l^{-1} in the aerobic reactor corresponding to nitrification.

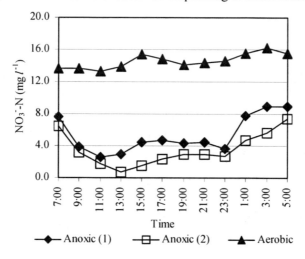

Figure 4.24. NO_3^--N concentration profiles of the three reactors of the laboratory-scale MLE activated sludge process in the optimization study.

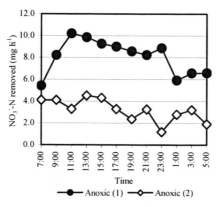

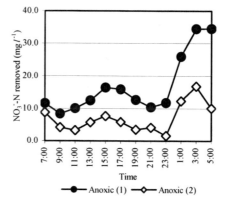

Figure 4.25(a). NO_3^--N removed, in terms of mass removal rate, in the two anoxic reactors of the laboratory-scale MLE activated sludge process in the optimization study.

Figure 4.25(b). NO_3^--N removed, in terms of influent-based concentration, in the two anoxic reactors of the laboratory-scale MLE activated sludge process in the optimization study.

Diurnal NO_3^--N removals in the first and second anoxic reactors, expressed in terms of mass removal rate and influent-based concentration, were calculated E 3.1 and 3.2, and are shown in Figures 4.25(a) and 4.25(b) respectively. In the first anoxic reactor, the denitrification rate reached a maximum of 10.2 mg NO_3^--N h^{-1} at about 11:00 during the morning peak period when the supply of COD in the feed was more than any other times, and a minimum of 5.4 mg NO_3^--N h^{-1} at about 07:00 during the base flow period, when COD in the feed was limited. Over a period of 24 h, a total of 193 mg NO_3^--N was removed in the first anoxic reactor. In the second anoxic reactor, the denitrification rate reached a maximum of 4.5 mg NO_3^--N h^{-1} at about 11:00 and a minimum of 1.2 mg NO_3^--N h^{-1} at about 23:00. Over a period of 24 h, a total of 77 mg NO_3^--N was removed, in which about 40% of that was accounted for by the first aerobic reactor. It clearly demonstrated the effect of anoxic volumetric ratio increase on denitrification. The maximum influent-based nitrate-nitrogen removal concentration was about 35 mg NO_3^--N l^{-1} during the base flow period, which was close to 32 mg NO_3^--N l^{-1}, the value obtained from sewage and plant denitrification potential analysis (Cao et al., 2003).

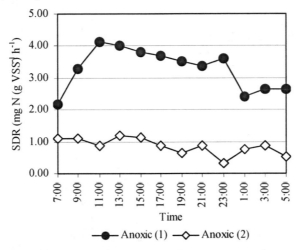

Figure 4.26. Specific denitrification rates in the two anoxic reactors of the laboratory-scale MLE activated sludge process in the optimization study.

Figure 4.26 shows the diurnal variations of the specific denitrification rate, which was obtained through the NO_3-N mass removal rate calculated by equation 3.1, divided by (MLVSS x V_{ANO}), in the first and second anoxic reactors. The general trend in the first anoxic reactor was that the specific rate reached a maximum of 4.1 mg NO_3^--N (g $VSS)^{-1}h^{-1}$ at about 11:00 during the morning peak period, and a minimum of 2.2 mg NO_3^--N (g $VSS)^{-1} h^{-1}$ during the base flow period. The specific rates in the first anoxic reactor were between those corresponding to the utilization of readily biodegradable COD (r_{DSS}) and rapidly hydrolysable biodegradable COD (r_{DX1}) (Table 2.4). The specific rate of denitrification in the second anoxic reactor reached a maximum of 1.2 mg NO_3^--N (g $VSS)^{-1}h^{-1}$ at about 13:00 and a minimum of 0.35 mg NO_3^--N (g $VSS)^{-1}h^{-1}$ at about 23:00. These two rates were between those corresponding to the utilization of rapidly hydrolysable biodegradable COD (r_{DXS1}) and endogenous respiratory COD (r_{Dend}), respectively (Table 2.4). These specific denitrification rates were in a similar range to that of the full-scale activated sludge process of Phase IV of Bedok WRP (Section 3.3.5.1).

4.3.4.2 Denitrification in the final clarifier

As shown in Figure 4.27, the NO_3^--N concentrations of RAS were lower than those at the inlet and outlet of the final clarifier, indicating the occurrence of denitrification in the final clarifier. Figure 4.28 shows that NO_3^--N removal rates calculated by Equations 3.1 and 3.2 reached a maximum of 2.0 mg NO_3^--N (g VSS)$^{-1}$h^{-1} at about 15:00 when the NO_3^--N concentration at the clarifier inlet reached a peak and a minimum of 0.04 mg NO_3^--N (g VSS)$^{-1}$ h^{-1} at about 11:00 when the NO_3^--N concentration at the clarifier inlet was minimal. Over a period of 24 h, a total of 20 mg NO_3^--N was removed. Given that the average NO_3^--N concentration of the final effluent was 14.1 mg NO_3-Nl^{-1} (Section 4.3.6; Table 4.9), the average overall NH_4^+-N based denitrification efficiency, which was calculated by Equation 3.6, was 52.3%.

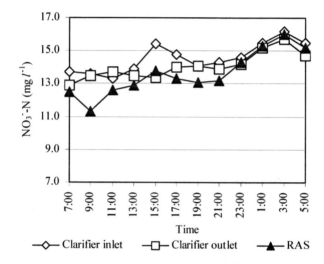

Figure 4.27. NO_3^--N concentration profiles at the inlet and outlet of the clarifier and of the RAS of the laboratory-scale MLE activated sludge process in the optimization study.

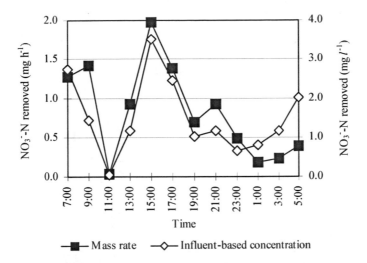

Figure 4.28. NO3--N removed, in terms of mass removal rate and influent-based concentration in the clarifier of the laboratory-scale MLE activated sludge process in the optimization study.

4.3.5 pH

The average pH in the second anoxic reactor was 6.78, which was 0.1 units higher than that in the first anoxic reactor (Figure 4.29). Between 13:00 and 06:00, the pH in the second anoxic reactor was mathematically significantly higher than that in the first anoxic reactor although no apparent differences were recorded between 07:00 and 12:00 when the pH in the first anoxic reactor was increasing as a result of peak mass flow of alkalinity. The average diurnal pH in the aerobic reactor was 6.14.

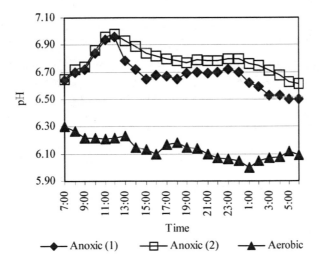

Figure 4.29. pH profiles of the three reactors of the laboratory-scale MLE activated sludge process in the optimization study.

4.3.6 Effluent quality

Figure 4.30 shows that the concentrations of NH_4^+-N and soluble TKN were higher during the daytime when the flow was higher. Table 4.9 shows the quality data of a 24-h composite sample of the final effluent. The alkalinity was lower than the expected value.

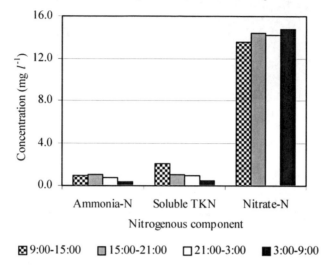

Figure 4.30. Nitrogenous component concentrations of the final effluent of the laboratory-scale MLE activated sludge process in the optimization study during various time intervals.

Table 4.9. Water quality (mg l^{-1}) of a 24-h composite sample of the final effluent of the optimized MLE activated sludge configuration.

Parameter	COD	SCOD	NH_4^+-N	NO_3^--N	Sol. TKN	ALK (as $CaCO_3$)
Concentration	39.2	20.5	0.84	14.1	1.33	17.7

As residual carbon in the secondary effluent affects the eventual quality of NEWater, diurnal effluent samples were also analyzed for soluble TOC and compared with data of the final effluent of Phase III of Kranji WRP (Figure 4.31). The average soluble TOC value of the laboratory effluent samples and that of the effluent of Phase III of Kranji WRP were 11.7 mg l^{-1} and 11.8 mg l^{-1}, respectively, and the difference was insignificant. This negligible difference illustrated the effectiveness of using this laboratory-scale system to investigate the full-scale process despite the differences in the anoxic volumetric ratios.

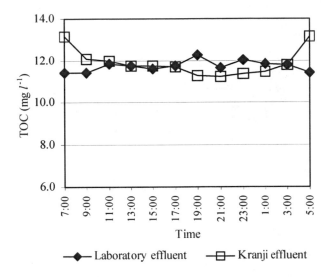

Figure 4.31. Diurnal soluble TOC values of the final effluent of the laboratory-scale MLE activated sludge process in the optimization study and Phase III of Kranji WRP.

4.3.7 Mass balance and yield coefficients

4.3.7.1 COD balance

The calculation was carried out using the same approaches as in Section 3.3.8.1. Table 4.10 summarizes the data calculated on a 24-h basis.

Table 4.10. Distribution of the COD (mg COD d^{-1}) in the laboratory-scale MLE activated sludge process in the optimization study.

Denitrified COD in the 1st anoxic reactor	Denitrified COD in the 2nd anoxic reactor	Denitrified COD in the final clarifier	Aeration Removed COD (RO_H)	Anabolized COD_{WAS}	COD in final effluent		Influent COD
					Soluble	Particulate	
555	221	57	1 980	1 461	287	281	4 841

* Calculated according to a SRT of 10 d and 10% COD content of the VSS

Figure 4.32 shows the percentage distribution of COD in the laboratory-scale MLE activated sludge process in the optimization study based on the values of Table 4.10: 30%, COD assimilation into biomass; 41%, oxygen-COD consumption in the aerobic reactor; 11%, COD removal in the first anoxic reactor; 5%, COD removal in the second anoxic reactor; 6%, soluble COD in the final effluent; 6%, particulate COD in the final effluent; and 1%, COD removal in the final clarifier. The COD assimilation in biomass and COD in the final effluent were close to those of the full- and laboratory-scale process of Bedok WRP. The sum of the COD removals in the anoxic reactors of 16% was higher than the full-scale data of 9% of Bedok WRP (Figure 3.20), but was almost similar to the laboratory-scale simulation data (18%) of Phase IV of Bedok WRP, where the high percentage was attributed to acetate addition into the sewage feed (Figure 4.16; Section 4.2.7.1).

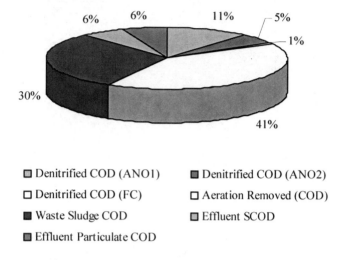

Denitrified COD (ANO1)   Denitrified COD (ANO2)
Denitrified COD (FC)   Aeration Removed (COD)
Waste Sludge COD   Effluent SCOD
Effluent Particulate COD

Figure 4.32. Percentage distribution of the influent COD in the laboratory-scale MLE activated sludge process in the optimization study.

The net sludge yield, calculated by Equation 3.7, and total catabolism yield, calculated by Equation 3.8, were 34.2% and 65.8%, respectively. The process oxygen coefficient, calculated by dividing RO_H by COD_{REM}, was 0.46 g O_2 (g COD_{REM})$^{-1}$. Compared with the total catabolism yield of 65.8%, the oxygen demand reduction credited to denitrification in the anoxic reactor was about 30% [(65.8-46)/65.8] of the COD catabolized, which was higher than that of the full-scale process of Phase IV of Bedok WRP (26%). The oxygen reduction based on total oxygen requirement was 16%, which was slightly higher than that of the full-scale system of 12% (Figure 3.20).

4.3.7.2 Nitrogen balance

Table 4.11 summarizes the nitrogen mass balance data on a 24-h basis. The total nitrogen input to the system (647 mg N) was the sum of the total nitrogen removed in the two anoxic reactors and final clarifier, that assimilated into cells (wasting), and that discharged in the effluent. This total nitrogen input was only 2% higher than the influent TKN load of the system.

Table 4.11. Distribution of the nitrogen (mg N) in the laboratory-scale MLE activated sludge process in the optimization study.

Nitrogen removed in 1st anoxic reactor	Nitrogen removed in 2nd anoxic reactor	Nitrogen removed in final clarifier	Nitrogen in wasted sludge (WAS)*	Nitrogen in final effluent				Calculated influent TN
				NH_4^+-N	NO_3^--N	Sol. Org-N	Susp. TKN**	
193	77	20	121	12	195	7	22	647

* Calculated according to a SRT of 10 d and 10% nitrogen content of the VSS
** Calculated from the TSS in the effluent and 10% nitrogen content of the VSS

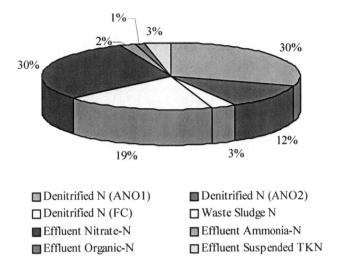

□ Denitrified N (ANO1)   ■ Denitrified N (ANO2)
□ Denitrified N (FC)   □ Waste Sludge N
■ Effluent Nitrate-N   ▨ Effluent Ammonia-N
▨ Effluent Organic-N   □ Effluent Suspended TKN

Figure 4.33. Percentage distribution of the influent nitrogen in the laboratory-scale MLE activated sludge process in the optimization study.

Figure 4.33 shows the percentage distribution of nitrogen in the activated sludge process. The percentage distribution in descending order was: 30%, NO_3^--N removal in the first anoxic reactor; 30%, NO_3^--N in the final effluent; 19%, nitrogen assimilation into sludge cells; 12%, NO_3^--N removal in the second anoxic reactor; 3%, suspended TKN in the final effluent; 3%, NO_3^--N removal in the final clarifier; 2%, NH_4^+-N in the final effluent; and 1%, soluble organic nitrogen in the final effluent. The NO_3^--N removal in the first and second anoxic reactor was 42%, which was higher than that of the full-scale process of Phase IV of Bedok WRP (26%) (Figure 3.21).

Table 4.12 records the nitrification and denitrification efficiencies calculated based on total nitrogen and ammonia-nitrogen by using Eqations 3.9 and 3.10. The total nitrogen-based denitrification efficiencies in the first and second anoxic reactors were 39.8% and 15.9%, respectively, which, in total, was 8.9% higher than 37.3%, the efficiency of the full-scale activated sludge process of Phase IV of Bedok WRP where the anoxic volume ratio was 25% of the total volume (Section 3.3.8.2). This also illustrated that denitrification was enhanced by increasing the volumetric ratio of the anoxic zone.

Table 4.12. Nitrification and denitrification efficiencies (%) of the laboratory-scale MLE activated sludge process in the optimization study.

Nitrification		Denitrification in the two anoxic reactor (based on total nitrogen)		Total denitrification	
Based on NH_4^+-N	Based on total nitrogen	First anoxic reactor	Second anoxic reactor	Based on NH_4^+-N	Based on total nitrogen
97.4	92.2	39.8	15.9	52.3	64.5

4.4 SUMMARY

Scale-down principle for activated sludge process

Three categories of ruling regimes for the scale–down of the activated sludge process are defined in this study:

i   feed conditions, including diurnal hydraulic flow, sewage compositions based on both conventional parameters and COD fractions;

ii   bioreactor system, including hydraulic flow pattern either ideal or non–ideal, configuration and sizes etc.; and

iii biochemical environment, including SRT, volumetric ratios of anaerobic/anoxic and aerobic zones in the activated sludge process; MLVSS concentrations, wasting rates, recycling ratios, HRT, temperature, DO concentrations in the systems.

The laboratory-scale process should share similar ruling regimes as those of the full-scale process investigated.

Simulation of the activated sludge process of Phase IV of Bedok WRP

The results of the laboratory-scale studies indicated that the bulk of the SCOD and nitrate nitrogen reduction could have occurred in the anoxic reactor. Ammonia-nitrogen oxidation occurred mainly in the first aerobic reactor, and the contribution of the second aerobic reactor was marginal. The SCOD, NH_4^+-N, and NO_3^--N concentration profiles were quite similar to those of the full-scale activated sludge process of Bedok WRP. The ranges of specific rates of nitrification and denitrification were similar to those of the full-scale activated sludge process of Bedok WRP.

The SCOD, NH_4^+-N, NO_3^--N and alkalinity of the effluent of the laboratory-scale activated sludge process were 12 mg SCOD l^{-1}, 0.5 mg NH_4^+-N l^{-1}, 9.0 mg NO_3^--N l^{-1} and 32.7 mg (as $CaCO_3$) l^{-1}, which were quite comparable to the corresponding values of 11 mg SCOD l^{-1}, 1.6 mg NH_4^+-N l^{-1}, 8.9 mg NO_3^--N l^{-1} and 37.0 mg (as $CaCO_3$) l^{-1} of the effluent of the full-scale activated sludge process. The percentage of COD and nitrogen assimilated into the biomass and the COD and nitrogen constituents in the effluent were comparable with those of the full-scale process of Phase IV of Bedok WRP.

The comparison of the results of the laboratory experiment with the full-scale system performance illustrated that the scale-down principle and the 'ruling regimes' of the activated sludge process defined in this study were workable. Nowadays, due to the application of automation technology, establishing a laboratory-scale activated sludge system designed based on the principles developed in this study is not expensive. Such laboratory-scale process experimentation provides a cost-effective tool in predicting full-scale system performance, optimizating the existing process, and developing new processes.

Optimization of the BNR activated sludge process in warm climates

Under an aerobic SRT of 3.75 d and with the anoxic volume increased from 25% to 62.5% (corresponding to 6.25 d anoxic SRT) of the total tank volume and feed conditions of Phase IV of Bedok WRP, the laboratory-scale MLE process showed complete nitrification, which demonstrated that the aerobic SRT and volume adopted in the existing activated sludge processes could be reduced significantly.

Denitrification efficiency in the expanded anoxic reactors was 55.7% when the anoxic volumetric ratio was 62.5%. It was higher than 37.3%, the denitrification efficiency in the anoxic compartment of the full-scale activated sludge process of Bedok WRP where the anoxic volumetric ratio was 25%. The oxygen requirement reduction was 16% compared with 12%, the data of the full-scale system with 25% anoxic volumetric ratio. Also, increases in average pH and alkalinity in the aerobic reactor of 0.2 units and 10 mg as $CaCO_3$ l^{-1}, respectively, were recorded. These results demonstrated that denitrification in the existing WRPs could be further enhanced by expanding the anoxic zone with the volume increase realized through reduction of the aerobic zone.

Laboratory optimization studies illustrated that in a warm climate, nitrification could be achieved at an aerobic SRT of 3.75 d, and denitrification can be greatly enhanced by enlarging the existing anoxic volumetric ratio from 25 to 50% or more of the total volume.

REFERENCES

Burrows L. J., Stokes A. J., West J. R., Forster C. F. and Martin A. D. (1999) Evaluation of Different Analytical Methods for Tracer Studies in Aeration Lanes of Activated Sludge Plants. Wat. Res. **33**(2), 367-374.

Burrows L. J., West J. R., Forster C. F. and Martin A (2001) Mixing Studies in An Orbal Activated Sludge System, Wat. SA **27**(1),79-83.

Cao Y.S., Ang C.M. and Zhao W. (2003) Performance Analysis of Full-and Laboratory-Scale Activated Sludge Process at Kranji and Bedok Water Reclamation Plants. Technical Report, Ref no. SUI/2001/030/TRTP1.

Cao Y. S., Raajeevan K. S. and Ang C. M. (2004) Optimization of Biological Nitrogen Removal in Municipal Wastewater Treatment by the Activated Sludge Process in Singapore. Technical Report, Ref no. SUI/2001/030/TRTP2.

Cui Yiqing (1997) Fungal Fermentation: Technological Aspect, PhD thesis, Delft University of Technology, The Netherlands.

EPA (1993) Manual Nitrogen Control, EPA/625/R-93/010, Washington D.C.

Goldstein R. A., Bourquin Al. W., Federle T. W., Grady C.P. L. and Mahaffey W. D. (1991) Environmental Biotechnology-From Flask to Field: A Review, In Environmental Biotechnology for Waste Treatment, (eds. G.S. Sayler *et al.*), Plenum Press, New York.

Grady C.P.L. Jr. (1993) Environmental Biotechnology: From Flask to Field. In Advanced Course on Environmental Biotechnology. Delft University of Technology, May 6-14, 1993, Delft, The Netherlands.

Henze M., Harremoës P., Janseen J. and Arvin E. (1997) Wastewater Treatment: Biological and Chemical Process, 2nd ed., Springer, Berlin.

Johnstone R.E. and Thring M.W. (1957) Pilot Plant, Model and Scale-Up in Chemical Engineering. McGraw Hill, New York.

Kossen N.W.F and Oosterhuis N.M.G. (1985) Modeling and Scaling-Up of Biorectors. In Biotechnology, (eds. Rehm H.J. and Reed G.), Vol 2, 572-605, VCH Verlaggesellschaft, Weinheim.

Kjellstrand R., Mattsson A., Niklasson C. and Taherzadeh M. J. (2005) Short Circuiting in a Denitrifying Activated Sludge Tank, Wat. Sci. Tech. **52**(11/12), 79-87.

Levenspiel O. (1972) Chemical Reactor Engineering, Prentice-Hall, Englewood Cliffs, New Jersey, USA.

Luyben K. Ch. A. M. (1993) Scale-Up/Scale-Down of (environmental) Biotechnological Process Using Regime Analysis. In Advanced Course on Environmental Biotechnology. Delft University of Technology, May 6-14, 1993, Delft, The Netherlands.

Metcalf and Eddy Inc. (2003) Wastewater Engineering: Treatment, Disposal and Reuse, 4th ed., McGraw-Hill, Washington, USA.

Oosterhuis N.M.G. (1984) Scale-Up of Bioreactor: A Scale-Down Approach, PhD Thesis, Delft University of Technology, The Netherland.

Rittmann B. and McCarty P. (2002) Environmental Biotechnology: Principles and Applications, International edition, McGraw-Hill, New York, USA.

Schmidtke N. M and Smith D.W. (1983) Scale-Up of Water and Wastewater Treatment. Butterworth Publisher, Durban.

Siegrist, H and Tschui, M. (1992). Interpretation of Experimental Data with Activated Sludge Model No. 1 and Calibration of the Model for Municipal Wastewater Treatment Plants. Wat. Sci. Tech. **25**(6), 167-183.

Stanier R., Ingraham J., Wheelis M. and Painter P. (1986) General Microbiology, 5th ed., Prentice-Hall, Englewood Cliffs, New Jersey, USA.

Sweere A. (1987) Response of Bakers' Yeast to Transient Environmental Conditions Relevant to Large-Scale Fermentation Processes, PhD Dissertation, Delft University of Technology, The Netherlands.

5

Mathematical modeling and simulation

5.1 INTRODUCTION

This chapter presents the results of modeling of the BNR activated sludge process in a warm climate. Almost all the contents presented in the preceding chapters are involved in the modeling process. The chapter begins with a brief literature review of the models, especially Activated Sludge Model No. 1 (ASM No. 1), and their applications for municipal sewage treatment. After the introduction, the approaches adopted in this study are described, including the development of the COD-based influent model and parameter identification, for which the measured data obtained from the laboratory experiment was adopted for parameter calibration, and both the measured data of the laboratory- and full-scale system performances was adopted for parameter verification. Simulation of the BNR activated sludge process in a warm climate using the verified ASM No. 1 is then presented. The chapter ends with recommendations on the design and upgrade of the BNR activated sludge process in warm climates.

5.1.1 Models of activated sludge process for municipal sewage treatment

Mathematical models used to describe activated sludge processes for municipal wastewater treatment were first developed in the 1960s, and only carbonaceous matter removal under steady-state conditions was considered (McKinney and Ooten, 1969), while nitrification was included in modeling later (Marais and Ekama, 1976). From the early 1980s, models describing both nitrification and denitrification with various kinetic parameters and stoichiometric coefficients were developed (Dold *et al.*, 1980; van Haandel *et al.*, 1981). In 1987, a Specialist Task Group of the International Water Association (IWA, formerly IAWPRC) published Activated Sludge Model No. 1 (ASM No. 1) (Henze *et al.*, 1987), in which the influent COD was fractionated into several constituents according to the biodegradation rate and physical form, and both the carbonaceous and nitrogenous matter removals were covered. The prevention of eutrophication motivated the development of models for phosphorus removal in addition to nitrogen and ASM No. 2 (Henze *et al.*, 1995) and ASM No. 2d (Henze *et al.*, 1999) are typical examples. The main modifications included the introduction of Phosphorus Accumulating Organisms (PAOs) and Glycogen Accumulating Organisms (GAOs), a further division of readily biodegradable COD into fermentable and fermentation products (volatile fatty acids), and substitution of lysis for decay. Later, anoxic and aerobic growths on the storage products were introduced in ASM No. 3 (Gujer *et al.*, 1999).

Initially, the researches and developments of activated sludge models focused on: (a) characterization of sewage for which various methods were developed (Grady, 1992; Vanrolleghem *et al.*, 1999); and (b) determination of stoichiometric coefficients and kinetic parameters for COD, nitrogen and phosphorus removals and other processes (Dold *et al.*, 1980; Henze *et al.*, 1987; Gujer *et al.*, 1999). For COD fractionation, no standard method has been formulated yet while for parameter determination, a number of model parameters were determined in laboratories or through site investigations. Many 'default values' are recommended for use under the specific conditions. However, the values of some parameters, such as half saturation and decay constants under different environmental conditions, are still uncertain.

Software packages for activated sludge modeling for various application purposes were developed. With the experience gained, the application of models and the modeling of activated sludge processes have become increasingly popular, and not just as educational tools. Modeling, as a professional tool, is playing important roles in investigating activated sludge processes for municipal sewage treatment for the following purposes:

i   to screen and pre-select process options and to conduct final design of municipal wastewater treatment processes;

ii   to diagnose the problems related to design or operation through study of the effects of process configurations and operation conditions to the performance and treatment efficiencies of the existing process and systems; and

iii   to optimize processes and operations through simulations of system performance under different processes, configurations and operation conditions.

Successful application of models and modeling has been reported in North America and Europe (Daigger and Nolasco, 1995; Ladiges *et al.* 1999; Carucci *et al.*, 1999; Makinia *et al.*, 2002; Salem *et al.* 2002; and van Veldhuizen *et al.* 1999). Several countries, especially those in Europe, are developing their own guidelines for easy and effective use of modeling in municipal wastewater treatment (Kruit and van Loosdrecht, 2002; Hulsbeek *et al.*, 2002; Roeleveld and van Loosdrecht, 2002). The approaches adopted in modeling play a decisive role in ensuring the reliability of the outcome; the critical steps are reliable feed input data and parameter identification. Due to its economic benefits and unique advantages, modeling is expected to draw more attention and, eventually, become a routine tool in municipal wastewater treatment work. However, there are still some factors that limit the application of modeling in the activated sludge process and these will be discussed in Section 5.1.3.

Table 5.1. Process kinetics and stoichiometry for multiple events in suspended growth cultures of ASM No. 1 (adopted from Henze et al., 1987).

Component → i	1	2	3	4	5	6	7	8	9	10	11	12	13	Process rate, r_j, $ML^{-3}T^{-1}$
j Process ↓	X_I	X_S	$X_{B,H}$	$X_{B,A}$	X_D	S_I	S_S	S_O[b]	S_{NO}	S_{NH}	S_{NS}	X_{NS}	S_{ALK}	
1 Aerobic growth of heterotrophs			1				$-\dfrac{1}{Y_H}$	$\dfrac{1-Y_H}{Y_H}$		$-i_{XB}$			$-\dfrac{i_{XB}}{14}$	$\hat{\mu}_H \left(\dfrac{S_S}{K_S+S_S}\right)\left(\dfrac{S_O}{K_{O,H}+S_O}\right) X_{B,H}$
2 Anoxic growth of heterotrophs			1				$-\dfrac{1}{Y_H}$		$-\dfrac{1-Y_H}{2.86\,Y_H}$	$-i_{XB}$			$\dfrac{1-Y_H}{14(2.86\,Y_H)} -\dfrac{i_{XB}}{14}$	$\hat{\mu}_H \left(\dfrac{S_S}{K_S+S_S}\right)\left(\dfrac{K_{O,H}}{K_{O,H}+S_O}\right)$ $\left(\dfrac{S_{NO}}{K_{NO}+S_{NO}}\right)\eta_g X_{B,H}$
3 Aerobic growth of autotrophs				1				$\dfrac{4.57-Y_A}{Y_A}$	$\dfrac{1}{Y_A}$	$-i_{XB}-\dfrac{1}{Y_A}$			$-\dfrac{i_{XB}}{14}-\dfrac{1}{7Y_A}$	$\hat{\mu}_A \left(\dfrac{S_{NH}}{K_{NH}+S_{NH}}\right)\left(\dfrac{S_O}{K_{O,A}+S_O}\right) X_{B,A}$
4 Death and lysis of heterotrophs		$1-f_D'$	-1		f_D'							$i_{XB}-f_D' i_{XD}$		$b_{L,H} X_{B,H}$
5 Death and lysis of autotrophs		$1-f_D'$		-1	f_D'							$i_{XB}-f_D' i_{XD}$		$b_{L,A} X_{B,A}$
6 Ammonification of soluble organic nitrogen										1	-1		$\dfrac{1}{14}$	$k_a S_{NS} X_{B,H}$
7 "Hydrolysis" of particulate organics		-1					1							$k_h \dfrac{X_S/X_{B,H}}{K_X + (X_S/X_{B,H})}\left[\left(\dfrac{S_O}{K_{O,H}+S_O}\right)\right.$ $\left.+ \eta_h\left(\dfrac{K_{O,H}}{K_{O,H}+S_O}\right)\left(\dfrac{S_{NO}}{K_{NO}+S_{NO}}\right)\right] X_{B,H}$
8 "Hydrolysis" of particulate organic nitrogen											1	-1		$r_7 (X_{NS}/X_S)$
Observed conversion rates, $ML^{-3}T^{-1}$														$r_i = \displaystyle\sum_{j=1}^{n} \Psi_{i,f_j}$

[a] All organic compounds (1–7) and oxygen (8) are expressed as COD; all nitrogenous components (9–12) are expressed as nitrogen.

[b] Coefficients must be multiplied by −1 to express as oxygen.

5.1.2 Models adopted and GPS-X

Three models were selected in this study: (1) ASM No. 1 to describe biological carbon and nitrogen removals; (2) a COD-based influent model; and (3) a secondary clarifier model capable of simulating settling and facilitating denitrification in the sludge blanket. Of the three, identification of parameters in ASM No. 1 and determination of the conversion coefficients of COD-based influent model were the main focuses, and, for this reason, a systematic and mechanistic approach was adopted.

5.1.2.1 Activated Sludge Model No. 1 (ASM No. 1)

Eight processes and thirteen components are accommodated in a matrix structure in ASM No. 1, as shown in Table 5.1 (Henze *et al.*, 1987). Practical experience of full-scale applications confirmed that the model structure is reliable and many default values, after temperature correction, are applicable.

Table 5.2 lists the typical values of model kinetic constants and stoichiometric coefficients applied to describe heterotrophic and autotrophic biodegradations (Henze *et al.*, 1987), and many of them are temperature dependent. However, little information is available with regards to suitable values of these parameters for application in warm climates. Hence, the selection and determination of the parameters that are applicable under Singapore's warm conditions became one of the major tasks of the modeling study.

Table 5.2. Typical parameter values at neutral pH (adopted from Henze *et al.*, 1987).

Symbol	Unit	Value at 20°C	Value at 10°C
	Stoichiometric parameters		
Y_A	g cell COD formed (g N oxidized)$^{-1}$	0.24	0.24
Y_H	g cell COD formed (g COD oxidized)$^{-1}$	0.67	0.67
f'_D	Dimensionless	0.08	0.08
$i_{N/XB}$	g N (g COD)$^{-1}$ in biomass	0.086	0.086
$i_{N/XD}$	g N (g COD)$^{-1}$ in endogenous mass	0.06	0.06
	Kinetic parameters		
$\hat{\mu}_H$	d^{-1}	6.0	3.0
K_S	g COD m^{-3}	20.0	20.0
$K_{O,H}$	g O_2 m^{-3}	0.20	0.20
K_{NO}	g NO_3^--N m^{-3}	0.50	0.50
b_H	day^{-1}	0.62	0.20
η_g	Dimensionless	0.8	0.8
η_h	Dimensionless	0.4	0.4
k_h	g slowly biodegradable COD (g cell COD)$^{-1}$ d^{-1}	3.0	1.0
K_X	g slowly biodegradable COD (g cell COD)$^{-1}$	0.03	0.01
$\hat{\mu}_A$	d^{-1}	0.80	0.3
K_{NH}	g NH_4^+-N m^{-3}	1.0	1.0
$K_{O,A}$	g O_2 m^{-3}	0.4	0.4
k_a	m^3 (g COD)$^{-1}$ d^{-1}	0.08	0.04

In the model, the COD and nitrogen components are characterized by their biodegradation rates and physical forms, which are different from conventional parameters such as COD, BOD and TSS, etc. Some typical values of the settled sewages of European countries (Henze *et al.*, 1987) and for Singapore are presented in Table 2.2.

5.1.2.2 COD-based influent model

New methods to characterize influent model state variables were able to improve the accuracy of the mathematical description. They affect the wide application of modeling as most of the measurement methods are time consuming and still under development, although report 99-WWF-3 (WERF, 2003) fulfilled a great need for standardization in the determination of the wastewater characteristics required for operation of simulation models. Therefore, to simplify the model application, an appropriate approach is to develop a set of conversion rules, which allows the use of the parameters of a regular monitoring programme in the influent input data file. Thus, a COD-based influent model was developed to satisfy this objective and is introduced in Section 5.2.

5.1.2.3 Secondary clarifier model

A secondary clarifier was included in the model process configuration, which recycles the return activated sludge (RAS) to the head of the activated sludge process. A one-dimensional non-reactive model (Simple 1d) (Hydromantis Inc., 1999) was adopted to describe thickening in the secondary sedimentation tanks even though sedimentation was not a focus of the study. In addition, to describe denitrification occurring in the sludge blanket, a CSTR following the clarifier was accommodated into the process configuration as adopted by Siegrist and Tschui (1992). Parameter identification of the clarifier was made to match the measured and simulated TSS concentrations of the effluent and RAS stream.

The CSTR kinetic parameters, corresponding to denitrification in the sludge blanket, were manipulated to obtain satisfactory matches between the measured and simulated NO_3^--N and NH_4^+-N concentrations in the RAS during the calibration and verification stages of the model. The measured sludge blanket volumes of laboratory-scale studies were used in the calibration and verification. However, it was recognized that due to the simplified conditions of the model compared to the complexity of hydraulic flow in the final clarifier, the model parameters of the sedimentation tank and CSTR would therefore have less mechanistic meaning, and thus obtaining a better fit between the measured and simulated concentration profiles became the main goal.

5.1.2.4 General Purpose Simulator – X (GPS–X)

Several software packages, programmed with complex differential equations and numerical approaches, are available in the market for application in wastewater treatment. Typical products include GPS-X (Hydromantis Inc., 1999), STOAT (WRc plc, 1998) and SIMBA (Kalker *et al.*, 1999).

The General Purpose Simulator - X (GPS–X), a product of Hydromantis Inc., Canada, is a comprehensive state-of-the-art software for dynamic modeling and simulation of municipal wastewater treatment plants. A wide variety of unit processes and unit operations commonly adopted in practice can be modeled with GPS-X. Due to its track record and powerful calculation capability, GPS–X (version 3.0) was used in the study.

5.1.3 Approaches of modeling and parameter identification

Figure 5.1 shows the general steps of a mathematical modeling procedure (Kossen and Oosterhuis, 1984). As mentioned in Section 5.1.2, three models were selected for this study. Input data of modeling municipal wastewater treatment plants is required as the next step, falling into the following categories:

i   influent wastewater characteristics, hydraulic flow, quality parameters, organic and nitrogen constituents for the influent model;

ii   plant physical data such as process flow sheet, system configuration and reactor dimensions, which determines the flow pattern;

iii   plant operating data such as DO concentration in wastewater, sludge wasting, ratios of MLR and RAS etc.;

iv   kinetic constants and stoichiometric coefficients of the process models; and

v   'measured data of process performance' i.e., the concentration profiles in individual compartments and the whole process, adopted in parameter identification.

To avoid 'rubbish in rubbish out', appropriate accuracy of these parameter/data values is one of the key factors in successful modeling.

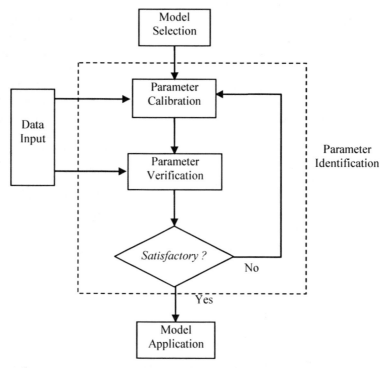

Figure 5.1. Steps of modeling (adopted from Kossen and Oosterhuis, 1985, Copyright © Wiley-VCH with permission).

Parameter identification, which includes calibration and verification, is the major work in the modeling procedure after selection of the appropriate models and data input. The availability of sufficient 'measured' data, which indicate the process performance, is a critical factor as, in calibration, the 'assumed values' of model parameters are adjusted to 'calibrated' values by minimizing the differences between the measured and simulated values. This is referred to as 'curve fitting'. In verification, a simulation is made with the calibrated models, and the outcomes are compared with 'another' set of measured data. Verification is deemed successful if the fit between the simulated and the measured data sets is satisfactory; if not, calibration and verification are repeated until a satisfactory match results.

Lack of sufficient data is another major obstacle that limits wide application of modeling. There may be difficulties in obtaining an appropriate set of influent data, as the sewage compositions used in modeling are different from conventional characterization. In addition to that, given the limited resources for monitoring, sampling and analysis, there is often insufficient 'measured' data on the system performance, which are needed for parameter identification. As a consequence, many parameters are 'artificially' manipulated to obtain a good curve fitting. This could lead to the outcomes losing their real meaning and applicability. In some cases, the measured data from the same activated sludge process operated with little or limited differences were adopted in both parameter calibration and verification and, consequently, the extrapolation of the outcomes could be limited and problematic.

Given the information collected, four approaches, which are applied in the parameter identification of modeling the activated sludge process in municipal wastewater treatment, are summarized below:

i   **Data from laboratory experimentation** - In the early years, most of the ASM No. 1 parameters were determined from laboratory-scale experiments. Henze *et al.* (1987) outlined some methods to measure kinetic constants such as $\hat{\mu}_A$ μ_A, K_{NH}, b_H, $\hat{\mu}_H$, K_S, etc. in the laboratory. Methods using oxygen uptake rate (OUR), nitrate uptake rate (NUR) and ammonia uptake rate (AUR) respirometry tests (Kappeler and Gujer, 1992; Spanjers, 1993; Cao and Alaerts, 1995; Vanrolleghem *et al.*, 1999) have become popular. In most cases, the sequencing batch reactor (SBR) is adopted. The kinetic constants obtained with these laboratory-scale facilities proved to be useful but, they might also be limited by the conditions of the experiment and the site, which might not be deemed similar to each other;

ii **Adoption of default value** - Recent experience indicates that, in many cases, many default values of parameters, including temperature-corrected coefficients, can describe and predict satisfactorily the performance of full-scale activated sludge processes (Daigger and Nolasco, 1995; Hulsbeek *et al.*, 2002). This approach is becoming increasingly popular as it features a marked reduction in the number of parameters to be identified, and is economically beneficial compared to laboratory experiment. Few reports are available on the applicability of this approach under warm conditions;

iii **Data from pilot-scale plant** - Calibration/verification was carried out by using measured data of pilot-scale facilities (Concha and Henze, 1992; Concha and Henze 1996; De La Sota *et al.*, 1994; Harromoës *et al.*, 1998). Confidence increases with regards to the reliability of the parameter values obtained as the performance of the pilot-scale process is much closer to that of the full-scale system. However, building and operating pilot-scale plants can be expensive especially when the effects of various alternatives in system configuration and operations have to be studied; and

iv **Data from full-scale plant** - A straightforward approach is to calibrate key parameters with the data obtained directly from full-scale plants (Lessouef *et al.*, 1992; Siegrist and Tschui, 1992). Ladiges *et al.* (1999) adopted this approach in modeling using ASM No. 1, and successfully applied the calibrated model in upgrading the wastewater treatment plant in Hamburg. Similar approaches were reported by Melcer (1999), Makinia *et al.* (2002), and Meijer (2004). For the latter, only three kinetic parameters were manipulated during calibration. However, in many cases, verification could be a problem due to a lack of an adequate number of measured data sets since it is costly to obtain a full set of measured data of a full-scale process.

5.1.4 Focuses and objectives

To pursue a cost-effective method to investigate BNR activated sludge processes, the approach for parameter identification of ASM No. 1 was formulated and designed with the following considerations and objectives:

i   explore the feasibility of identifying model parameters from the data of scaled-down laboratory experiment, which are applicable in describing the performance of the full-scale process. For this purpose, several sets of measured data obtained from laboratory-scale investigations were adopted in calibration, and the measured data of the full-scale investigations carried out in two different WRPs were used for verification;

ii   follow the strategy of adopting, to the maximum extent, the default parameter values directly or after temperature correction whenever possible, and calibrate only the parameters whose default values may not be applicable. This reduces the number of parameters to be identified and prevents, to a large extent, the 'uncertainty' in modeling (Chen, 1987) i.e., similar fit between the simulated and measured data may be achievable with several different combinations of parameter values of the ASM No. 1 and other AS models. Another principle is to optimize the parameter values within the rational ranges of their original meaning rather than only aiming at best fit between the simulated and measured concentration profiles;

iii   develop a simple COD-based influent model derived from the detailed sewage characterization. This will enable the use of regular monitoring data as influent input data in modeling under Singapore conditions;

iv   study the effects of influencing factors on the performance for the purpose of optimization and to investigate several issues which are not easy to study on site and in the laboratory, through the simulation of the BNR activated sludge process using ASM No. 1 with the verified parameters; and

v   develop a computer-aided capacity for formulating the design guidelines of the BNR activated sludge process, which are to be adopted in the upgrading of existing plants in warm climates.

5.2 DEVELOPMENT OF A COD-BASED INFLUENT MODEL

The influent COD of diurnal samples were fractionated according to ASM No. 1 in this project (Section 2.3.3.2), and the results were applied successfully as direct inputs to ASM No. 1 to model the activated sludge process (Raajeevan, 2003).

For this study a COD-based influent model in GPS-X (Hydromantis Inc., 1999) was selected, and derivation of conversion coefficients between the ASM No. 1 influent composition and the conventional parameters in the COD-based model was carried out.

Figures 5.2 and 5.3 show the relationships between the conventional parameters in the COD-based model and the COD fractions in ASM No. 1, in both soluble and particulate COD forms, respectively (X_U is non-biodegradable particulate COD due to cell decay).

The COD value determined from the filtrate of a 0.45 μm filter was assumed to be SCOD. It should be noted that this value was slightly overestimated as it may include part of the colloidal and fine particulates that passed through the filter (Henze et al., 1995). The S_I could be taken as the residual SCOD of the effluent of the activated sludge system, so the frsi (S_I/SCOD) could be calculated based on the measured effluent SCOD. According to the characterization studies (Table 2.2; Section 2.3.3.1), the S_I under Singapore conditions was taken as 20 mg COD l^{-1}, which was similar to reported values (Henze et al., 1987). Hence, the readily biodegradable COD (S_S) was calculated from the equation below:

$$S_S = SCOD - S_I$$

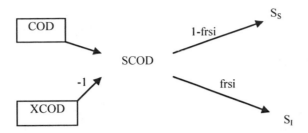

Figure 5.2. Influent soluble components and their relationships in the COD-based model (adopted from Hydromantis Inc., 1999 with permission).

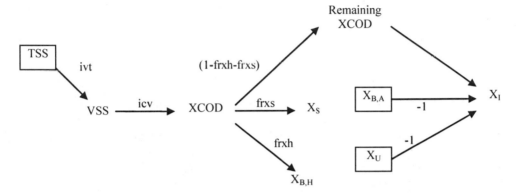

Figure 5.3. Influent particulate COD components and their relationships in the COD-based model (adopted from Hydromantis Inc., 1999 with permission).

In ASM No. 1, the biodegradable particulate COD is represented by a consolidated parameter, X_S (Henze *et al.*, 1992), while in the COD-based model, the components X_S, $X_{B,H}$ and $X_{B,A}$ are identified separately, similarly to that of ASM No. 2 (Henze *et al.* 1995). Figure 5.4 illustrates the discrepancies in the COD fractions in ASM No. 1 and the COD-based influent model, which assumed $X_{B,A}$ and X_U to be 1 mg COD l[-1] and non-detectable, respectively (Hydromantis 1999), and the related analytical values. To use the COD-based influent model, the conversion coefficients relating these conventional parameters, such as the COD, SCOD, TSS or VSS, with the ASM No. 1 components (state variables), have to be derived and calculated.

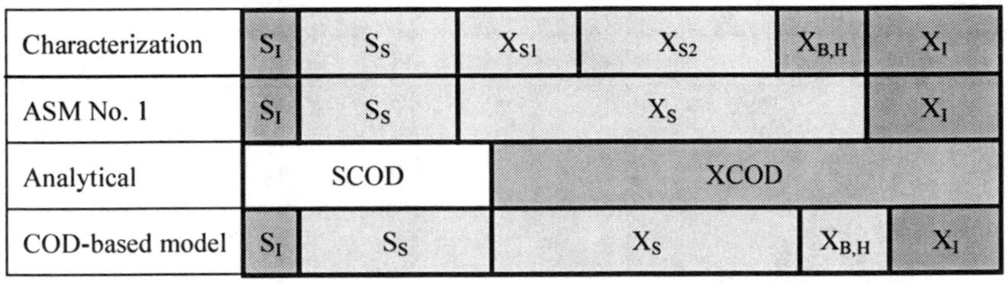

Characterization	S_I	S_S	X_{S1}	X_{S2}	$X_{B,H}$	X_I
ASM No. 1	S_I	S_S	X_S			X_I
Analytical	SCOD		XCOD			
COD-based model	S_I	S_S	X_S		$X_{B,H}$	X_I

Figure 5.4. COD fractionation in the COD-based influent model.

The reliabilities of sewage COD and SCOD measurements are higher than those of TSS and VSS. To avoid any errors resulting from TSS and VSS measurements, XCOD was, therefore, calculated from the COD and SCOD by the equation below, rather than by using the conversion factors ivt and icv (Figure 5.2).

$$XCOD = COD - SCOD$$

It should be noted that taking analytical SCOD instead of S_S in the calculation of XCOD and the sum of S_S, X_{S1} and X_{S2} as biodegradable COD (BCOD) can trade off the possible overestimation of X_I (Table 2.2; Section 2.3.3.1). The BCOD in the COD-based model is the sum of S_S, X_{S1}, and X_{S2} as shown in Figure 5.4. Of the two possible ways of obtaining X_S, either directly from the sum of X_{S1} and X_{S2} or from (BCOD-S_S), the latter was chosen to avoid overestimating the total BCOD since a portion of the X_{S1} is included in the S_S of the COD-based model.

The frxs ratio, defined as the fraction of X_S in the XCOD, is calculated in the COD-based influent model from:

$$frxs = \frac{X_S}{XCOD} = \frac{(BCOD - S_S)}{(COD - SCOD)}$$

To explore the relationship between BCOD and COD, the results of COD fractionations, derived from diurnal settled sewage samples (Section 2.3.3.2), were reviewed. Table 5.3 shows a summary of the data of the settled sewages of Bedok WRP and Phase III of Kranji WRP under different hydraulic conditions.

The COD fractions changed with hydraulic flow. Typically, the S_S was depleted and the total COD values were very low during the low flow period at night. However, the BCOD/COD ratios varied between 36 and 57%. For simplification, the BCOD/COD ratio of 50% was adopted and the conversion factor, frsi, was introduced for further simplification.

Table 5.3. COD fractions and frxs values calculated based on the data of Bedok and Kranji WRPs.

Parameter	S_S	SCOD	$\dfrac{S_S}{SCOD}$	$X_{S1}+X_{S2}$	COD	$\dfrac{X_{S1} + X_{S2}}{SCOD}$	$S_S+X_{S1}+X_{S2}$ (BCOD)	$\dfrac{BCOD}{COD}$
	(mg l^{-1})	(mg l^{-1})	(%)	(mg l^{-1})	(mg l^{-1})	(%)	(mg l^{-1})	(%)
				Bedok WRP				
Peak	66	173	38	147	512	29	213	42
Normal	51	131	39	53	286	19	104	36
Low flow	6	87	7	90	193	47	96	50
				Phase III of Kranji WRP				
Peak	36	149	24	208	430	48	244	57
Normal	50	127	39	108	369	29	158	43
Low flow	ND	119	ND	113	249	45	113	45

ND: non detectable

From the equations above, the frxs could be expressed in terms of the COD and SCOD using the following equation:

$$frxs = \frac{\{COD \times 50\% - (1 - frsi) \times SCOD\}}{(COD - SCOD)}$$

It must be noted that frsi is variable as the COD and SCOD values varied with the hydraulic flow as shown in Table 5.3.

The TSS and VSS concentrations were measured and then the conversion factor, ivt (VSS/TSS), was calculated. The analyzed value of the ivt under Singapore conditions was 96%. The value of the conversion factor, icv (XCOD/VSS), was calculated from the values of XCOD, TSS and ivt. For the frxh fraction (X_H/XCOD), the default value of 20% was used (Kappeler and Gujer, 1992). Also, the default values of 1 and 0 mg COD l^{-1} as suggested by Hydromantis Inc. (1999), were assumed for $X_{B,A}$ and X_U, respectively.

Figure 5.5 shows the nitrogenous constituents in the influent sewage and their relationships between ASM No. 1 and the COD-based influent model. The diurnal conversion factor, fnh (NH_4^+-N/TKN), was calculated from the measured NH_4^+-N and TKN values.

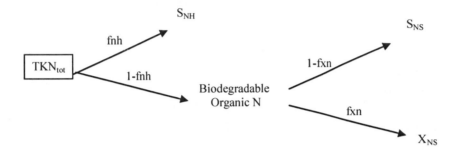

Figure 5.5. Influent nitrogenous components and their relationships in the COD-based model (adopted from Hydromantis Inc., 1999 with permission).

The fraction of particulate biodegradable organic nitrogen in the total organic nitrogen, fxn (X_{NS}/(TKN– NH_4^+-N)), was calculated from the characterization data and was 80% (Cao et al. 2004a). S_{NS} is the difference between soluble TKN and NH_4^+-N and is obtained from measured values. It should be emphasized that since all the nitrogen components in this model were assumed biodegradable, the simulated TN concentration in the final effluent could be less than the measured concentration by about 1.0 mg N l^{-1} due to the presence of recalcitrant soluble inorganic nitrogen in the influent (Henze et al., 1995; Cao et al. 2003).

The parameters, including the conversion factors, in the COD-based influent model adopted in this study are summarized in Table 5.4. As recommended in the literature (Henze et al., 1995; Hulsbeek et al., 2002), the values of these parameters were obtained from: (a) analytical measurements such as TSS, COD, SCOD, TKN and NH_4^+-N, etc.; (b) suggested values derived from characterization data; or (c) calculations from known equations and suggested methods (mainly for conversion factors).

According to Table 5.4, the conventional parameters analyzed in the regular monitoring/sampling programme such as COD, SCOD, TSS, VSS, TKN and NH_4^+-N together with the suggested and calculated coefficients are adequate for the COD-based influent model used in ASM No. 1 under Singapore conditions. This simplifies, largely, the characterization work and eases application of modeling and simulation.

Table 5.4. Parameters in the COD-based influent model under Singapore conditions.

Symbol	Description	Unit	Range/ Value	Method	Formula
Parameter					
TSS	Total suspended solids	mg l^{-1}	60-200	Analytical	-
VSS	Volatile suspended solids	mg l^{-1}	55-190	Analytical	TSS x ivt
COD	Chemical oxygen demand	mg COD l^{-1}	350-450	Analytical	-
SCOD	Soluble COD	mg COD l^{-1}	120-150	Analytical	-
XCOD	Particulate COD	mg COD l^{-1}	230-300	Calculated	(COD-SCOD)
S_S	Readily biodegradable COD	mg COD l^{-1}	-	Calculated	(SCOD-S_I)
X_S	Slowly biodegradable COD	mg COD l^{-1}	-	Calculated	XCOD x frxs
$X_{B,H}$	Heterotrophs	mg COD l^{-1}	-	Calculated	XCOD x frxh
$X_{B,A}$	Autotrophs	mg COD l^{-1}	1	Suggested	-
X_U	Unbiodegradable particulate COD from cell decay	mg COD l^{-1}	0	Suggested	-
S_I	Inert soluble organics	mg COD l^{-1}	20-30	Suggested	SCOD x frsi
X_I	Inert particulate organics	mg COD l^{-1}		Calculated	(XCOD-X_S-$X_{B,H}$)
TKN	Total Kjeldhal Nitrogen	mg N l^{-1}	40-52	Analytical	-
S_{NH}	Ionized ammonia nitrogen	mg N l^{-1}	30-40	Analytical	-
Org. Bio. N	Organic biodegradable nitrogen	mg N l^{-1}	-	Calculated	(TKN - S_{NH})
S_{NS}	Soluble organic biodegradable nitrogen	mg N l^{-1}	-	Calculated	Org Bio. N x (1-fxn)
X_{NS}	Particulate organic biodegradable nitrogen	mg N l^{-1}	-	Calculated	Org Bio. N x fxn
Conversion factor					
fxn	$\dfrac{X_{NS}}{Org.bio.N}$	%	80	Suggested	-
ivt	$\dfrac{TSS}{VSS}$	%	96	Suggested	-
icv	$\dfrac{XCOD}{VSS}$	%	-	Calculated	$\dfrac{(COD-SCOD)}{(TSS \times 96\%)}$
frsi	$\dfrac{S_I}{SCOD}$	%	15	Suggested	-
frxs	$\dfrac{X_S}{XCOD}$	%	-	Calculated	$\dfrac{(COD \times 50\% - SCOD \times 85\%)}{(COD-SCOD)}$
frxh	$\dfrac{X_{BH}}{XCOD}$	%	20	Suggested	-

5.3 PARAMETER CALIBRATION OF ASM NO. 1 WITH LABORATORY EXPERIMENTATION DATA

Biological nitrogen removal, comprising nitrification and denitrification, was the focus of parameter identification. The related kinetic constants and stoichiometric coefficients under warm conditions were investigated by comparing the simulated and measured NO_3^--N and NH_4^+-N concentration profiles in parameter identification. The SCOD and VSS concentrations were also studied but illustrated only in steady-state calibration. Different sets of measured data from both laboratory-scale and full-scale investigations of different WRPs were used in the parameter calibration and verification.

Upon completion of layout building and data input, definition of initial conditions was required prior to parameter calibration/estimation.

5.3.1 Definition of initial conditions and sequence for model calibration

The initial conditions adopted in the simulation are closely related to the outcomes of the simulation, and are particularly true for the effect of autotrophic nitrifier concentrations on nitrification efficiency such that it was adopted as a major indicator for justification.

Consistent with theory and experience, simulations showed that the performance of the activated sludge process reached a 'well-developed' state after a simulation period of three SRTs, provided that the diurnal mass loads and influent concentrations remained the same (Rittmann and McCarty, 2001). Therefore, the initial biomass concentrations were taken from the simulated values after a simulation period of three SRTs, as being default initial biomass concentrations for dynamic state.

On some occasions, the biomass concentrations after a simulation period of one SRT were used for model initialization after appropriate estimates of the autotrophic nitrifier and heterotrophic bacteria concentrations. The laboratory experiments proved that the system reached a 'pseudo' well-developed state under the above conditions (Cao *et al.*, 2003, Cao *et al.*, 2004b). The duration of the simulation for the dynamic state was selected according to the actual duration of the sampling programme.

The sequence adopted in parameter calibration is described below:

i   to achieve a solid balance through the comparison of measured and simulated VSS concentrations in the reactors, with the input for wasting according to the average SRT data and calibrating b_H when necessary;

ii   to achieve an appropriate fit between the simulated and measured NO_3^--N and NH_4^+-N concentration profiles in the RAS by modifying relevant parameters in the CSTR simulating denitrification in the sludge blanket;

iii   to calibrate the kinetic and stoichiometric parameters of denitrification such as μ_H and Y_{HD}, etc., that are related to the heterotrophic bacteria, by comparing the simulated and measured NO_3^--N concentration profiles in the anoxic reactor where denitrification was carried out by heterotrophic bacteria; and

iv   to calibrate the kinetic constants of nitrification such as μ_A and b_A, etc., that are related to the autotrophic nitrifiers, by comparing the simulated and measured NH_4^+-N concentration profiles in the aerobic reactor where nitrification was carried out by autotrophic nitrifiers.

It might be addressed that the curve fitting in (ii) might not be so 'satisfactory', as the models of secondary clarifier are relatively empirical rather than mechanistic.

In principle, one set of measured data is sufficient for parameter calibration/estimation. To study the possible differences in calibration by using the data measured under steady- and dynamic-state, two sets of measured data were adopted in calibration. The primary calibration was made with measured data obtained from the laboratory-scale studies under steady-state experimental conditions, while the secondary calibration was made with data obtained from the laboratory-scale studies under dynamic-state conditions.

5.3.2 Primary calibration: steady-state conditions

The laboratory-scale system used to determine the optimal ratio of MLR and designed to simulate the MLE configuration of Phase IV of Bedok WRP, with an anoxic volume ratio of 25% under steady-state feed flow conditions, was selected for primary calibration. The schematic diagram and the operation conditions of the configuration are shown in Figure 5.6. The model configuration built in GPS-X incorporated all the features illustrated in Figure 5.6 and a CSTR was introduced in the RAS recycle line to simulate denitrification in the clarifier sludge blanket.

The influent composition was obtained by averaging the influent data during the monitoring period, a similar approach as Concha and Henze (1992; and 1996) and Makinia *et al.* (2002). The concentrations of the parameters and the conversion factors in the COD-based influent model are presented in Table 5.5.

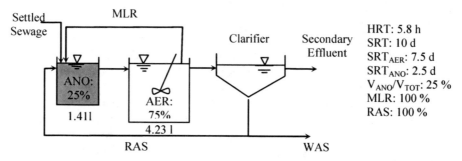

Figure 5.6. Schematic diagram of the laboratory-scale MLE process of Phase IV of Bedok WRP under steady state conditions.

Table 5.5. Influent composition and stoichiometric coefficients used in the primary calibration.

Parameter	Unit	Value in this study
COD	g COD m^{-3}	259
TSS	g m^{-3}	89
TKN	g N m^{-3}	44.8
NH$_4^+$-N	g N m^{-3}	32.1
SCOD	g COD m^{-3}	107
frsi	-	0.19
frxs	-	0.32
frxh	-	0.2
fnh	-	0.72
fxn	-	0.8
icv	g COD(g VSS)$^{-1}$	1.78
ivt	g VSS(g TSS)$^{-1}$	0.96

Table 5.6. Assumed and calibrated values of kinetic constants and stoichiometric coefficients.

Parameter	Unit	Temperature corrected coefficient (θ)*	Assumed	Calibrated
Kinetic Parameter				
$\hat{\mu}_H$	d^{-1}	0.069	12.00	12.00
K_S	g COD m^{-3}	0.0	20.00	20.00
b_H	d^{-1}	0.113	1.92	0.62
η_h	-	0.0	0.40	0.40
η_g	-	0.0	0.80	0.80
k_h	g SBCOD(g cell COD)$^{-1}$.d^{-1}	0.11	9.00	9.00
K_X	g SBCOD(g cell COD)$^{-1}$	0.11	0.09	0.09
k_a	m^3 (g COD)$^{-1}$d^{-1}	0.69	0.16	0.16
$\hat{\mu}_A$	d^{-1}	0.098	2.13	1.5
K_{NH}	g N m^{-3}	0.0	1.0	1.0
b_A	d^{-1}	-	0.15	0.15
$K_{O,H}$	g O$_2$ m^{-3}	0.0	0.20	0.20
$K_{O,A}$	g O$_2$ m^{-3}	0.0	0.40	0.40
K_{NO}	g N m^{-3}	0.0	0.50	0.50
Stoichiometric coefficient				
COD/VSS ratio	g COD(g VSS)$^{-1}$	-	1.48	1.48
VSS/TSS ratio	g VSS(g TSS)$^{-1}$	-	0.75	0.75
BOD$_5$/BOD$_U$ ratio	-	-	0.66	0.66
Aerobic Y_H	g cell COD(g COD)$^{-1}$	-	0.67	0.67
Anoxic Y_{HD}***	g cell COD(g COD)$^{-1}$	-	0.67	0.53**
$i_{N/XB}$	g N(g cell COD)$^{-1}$	-	0.086	0.086
$i_{N/XD}$	g N(g cell COD)$^{-1}$	-	0.06	0.06
f'_D	-	-	0.08	0.08
Y_A	g cell COD(g COD)$^{-1}$	-	0.24	0.24

* Derived from data, at 10 and 20°C, of ASM No. 1 (Henze et al., 1987) according to the equations $K_{30} = K_{20} \times e^{\theta(30-20)}$ and $\theta = (\ln K_{20} - \ln K_{10})/10$
** Values of parameters identified through calibration that were different from the default values at 30°C
*** Sludge blanket was also assumed to be anoxic

In the calibration, the DO concentration in the whole anoxic reactor was set at 0.05 mg l^{-1}, and that in the aerobic reactor was set according to the measured value (2-2.5 mg l^{-1}). The initial concentrations in the reactors were assumed to be similar to those of the measured average values.

Table 5.6 shows the kinetic constants and stoichiometric coefficients before and after the calibration. Most of the kinetic constants and stoichiometric coefficients were either adopted from default values or temperature-corrected according to the data at 10 and 20°C in ASM No. 1 where necessary, such as μ_A and k_h etc (Henze et al., 1987). Default values were used for saturation constants and switching constants such as K_S, K_{NH}, K_{NO}, $K_{O,H}$ and $K_{O,A}$

assuming that they were not temperature related. However, K_X, the saturation constant of hydrolysis, was corrected according to the correction coefficient in ASM No. 1. A value of 0.15 d^{-1} was assumed for b_A, which was the upper limit of the recommended range of ASM No. 1 (Henze *et al.*, 1987).

There were three parameters ($\hat{u}_A$, b_H, and anoxic Y_H) whose values were different from those assumed initially after calibration. The value of $\hat{u}_A$ reduced to 1.5 d^{-1} corresponding to the extrapolated value at 26.5°C after temperature correction as suggested in ASM No. 1. The calibrated b_H value of 0.62 d^{-1} of the activated sludge process was the default value at 20°C (Henze *et al.*, 1987) but lower than that assumed initially. This was in agreement with practical modeling experience i.e., b_H value of 0.62 d^{-1} is quite stable (Takàcs, 2006) and with those reported by Kappeler and Gujer (1992) and Henze *et al.* (1995) where b_H at 22°C varies between 0.1 and 0.4 d^{-1} and a lower value of 0.4 d^{-1} for the lysis constant of heterotrophs in ASM No. 2, respectively. However, b_H of the CSTR, accounting for denitrification of the final clarifier, was 1.62 d^{-1}, though the other kinetic constants and stoichiometric coefficients describing the CSTR were the same as those of the anoxic reactor. The calibrated value of Y_H in the anoxic reactor was 0.53 g cell COD (g COD)$^{-1}$, which was lower than the default value of 0.67 g cell COD (g COD)$^{-1}$ (Henze *et al.*, 1987). In fact, some reports have shown that the yield coefficient under anoxic condition was markedly lower than under aerobic condition (Orhon *et al.*, 1996).

Figures 5.7(a)-(f) show the measured and simulated data of SCOD, VSS, NH$_4^+$-N and NO$_3^-$-N concentrations in the anoxic, aerobic reactors and RAS. The satisfactory fit between the simulated and measured VSS concentrations in Figures 5.7(a), (c) and (e) showed that the solid balance was met and, thus, the value selected for b_H was appropriate. It was also noted that the measured VSS/TSS ratio was very close to the assumed default and site ratio of 0.75.

The simulated average SCOD and NO$_3^-$-N concentrations in the anoxic reactor were close to the measured average values (Figures 5.7(a) and (b)) indicating that the calibrated kinetic constants and stoichiometric coefficients of the heterotrophs were suitable for describing denitrification under the laboratory-scale conditions.

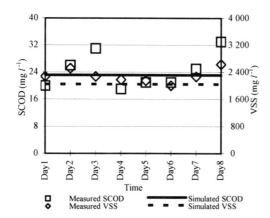

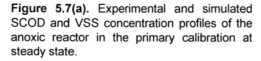

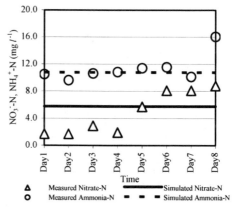

Figure 5.7(a). Experimental and simulated SCOD and VSS concentration profiles of the anoxic reactor in the primary calibration at steady state.

Figure 5.7(b). Experimental and simulated NH$_4^+$-N and NO$_3^-$-N concentration profiles of the anoxic reactor in the primary calibration at steady state.

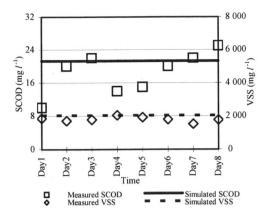

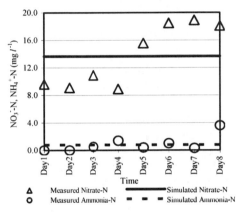

Figure 5.7(c). Experimental and simulated SCOD and VSS concentration profiles of the aerobic reactor in the primary calibration at steady state.

Figure 5.7(d). Experimental and simulated NH_4^+-N and NO_3^--N concentration profiles of the aerobic reactor in the primary calibration at steady state.

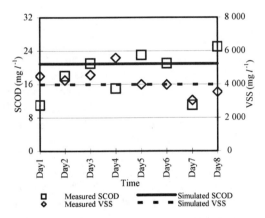

Figure 5.7(e). Experimental and simulated SCOD and VSS concentration profiles of the RAS in the primary calibration at steady state.

Figure 5.7(f). Experimental and simulated NH_4^+-N and NO_3^--N concentration profiles of the RAS in the primary calibration at steady state.

The differences between the simulated and measured SCOD values were within the accepted range of measurement errors in an engineering context (Figure 5.7(c)). The simulated average value of NH_4^+-N in the aerobic reactor was 0.2 mg NH_4^+-N l^{-1}, which was of the same order of magnitude as the measured average value of 0.9 mg NH_4^+-N l^{-1} (Figure 5.7(d)). This illustrated that the calibrated kinetic constants and stoichiometric coefficients of the autotrophic nitrifiers were suitable for describing the nitrification process.

Figures 5.7(e) and (f) show the simulated and measured SCOD, VSS, NH_4^+-N and NO_3^--N concentrations in the RAS. These data fitted satisfactorily with each other. However, the selected b_H value of 1.62 d^{-1} for the CSTR, describing denitrification in the sludge blanket, was higher than those of the anoxic and aerobic reactors.

Table 5.7 summarizes the average values of and the differences between the measured and simulated VSS, SCOD, NH_4^+-N and NO_3^--N concentrations in the anoxic, aerobic reactors and the RAS. The differences were within the range of reported values (Concha and Henze, 1992; Concha and Henze, 1996) and the accepted range of measurement error in an engineering context. Simulations were also made with the calibrated model at steady-state with 50 and 200% MLR ratios. The differences in these sets were also within the same range as those of Table 5.7 and compared well with the measured data (Cao *et al.*, 2004c). Thus, these comparisons consolidated the validity of the kinetic constants and stoichiometric coefficients of the calibrated model.

Table 5.7. Measured and calibrated average parameter concentrations (mg l^{-1}) and their differences in the reactors and the RAS of the laboratory-scale MLE configuration at steady state.

Value	VSS			SCOD			NH_4^+-N			NO_3^--N		
	ANO	AER	RAS	ANO	AER	RAS	ANO	AER	RAS	ANO	AER	RAS
Measured	2 284	1 799	4 178	25	19	18	11.3	0.9	0.9	4.9	16.5	13.7
Calibrated	2 045	2 037	3 987	23	21	21	10.8	0.2	0.8	5.8	18.1	13.6
Difference	239	-238	191	2	-2	-3	0.6	0.7	0.1	-0.9	-1.6	0.1

5.3.3 Secondary calibration: dynamic-state conditions

The kinetic constants and stoichiometric coefficients calibrated with the measured data at steady-state (Table 5.6) were further tested with the measured data obtained from a laboratory-scale experiment that simulated the performance of the MLE activated sludge process of Phase IV of Bedok WRP under diurnal feed conditions (Section 4.2.1). Figure 5.8 shows the schematic diagram of the laboratory-scale configuration with the key process parameters and operation conditions. The details of the experimental conditions and the performance results were introduced in Section 4.2.1 and reported in Cao *et al.*, (2004b). The hydraulic peak flow occurred between 09:00 and 13:00 (Figures 4.1 and 4.2) when most of the conventional parameter concentrations were at their highest (Table 4.1). The low (base) flow occurred between 01:00 and 07:00. Table 5.8 presents the feed flow, compositions and conversion factors used in the COD-based influent model.

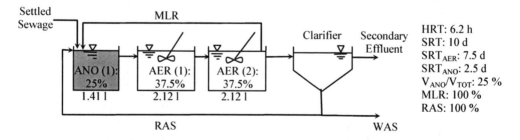

Figure 5.8. Schematic diagram of the laboratory-scale MLE system of Phase IV of Bedok WRP under dynamic state conditions.

Table 5.8. Diurnal influent data of the laboratory-scale MLE process of Phase IV of Bedok WRP under dynamic state conditions. *q: flow rate corresponding to total effective bioreactor volume of 1 000 m³.

Time	q (m³d⁻¹)	COD (mgl⁻¹)	SCOD (mgl⁻¹)	BOD₅ (mgl⁻¹)	TSS (mgl⁻¹)	NH₄⁺-N (mgl⁻¹)	TKN (mgl⁻¹)	ALK (mgl⁻¹)	Sₛ (mgl⁻¹)	Xₛ (mgl⁻¹)	ivt -	icv -	frsi -	frxs -	fnh -	fxn -
07:00	3 265	230	125	73	68	28.1	32.0	168.8	105	32.6	0.96	1.61	0.16	0.31	0.88	0.8
08:00	4 081	230	125	73	68	28.1	32.0	168.8	105	32.6	0.96	1.61	0.16	0.31	0.88	0.8
09:00	4 898	230	125	73	68	28.1	32.0	168.8	105	32.6	0.96	1.61	0.16	0.31	0.88	0.8
10:00	5 689	280	146	82	85	35.1	40.0	193.3	125	57.6	0.96	1.64	0.14	0.43	0.88	0.8
11:00	5 689	280	146	82	85	35.1	40.0	193.3	125	57.6	0.96	1.64	0.14	0.43	0.88	0.8
12:00	5 689	280	146	82	85	35.1	40.0	193.3	125	57.6	0.96	1.64	0.14	0.43	0.88	0.8
13:00	5 689	280	146	82	85	35.1	40.0	193.3	125	57.6	0.96	1.64	0.14	0.43	0.88	0.8
14:00	5 204	230	125	73	68	28.1	32.0	168.8	105	32.6	0.96	1.61	0.16	0.31	0.88	0.8
15:00	4 719	230	125	73	68	28.1	32.0	168.8	105	32.6	0.96	1.61	0.16	0.31	0.88	0.8
16:00	4 719	230	125	73	68	28.1	32.0	168.8	105	32.6	0.96	1.61	0.16	0.31	0.88	0.8
17:00	4 719	230	125	73	68	28.1	32.0	168.8	105	32.6	0.96	1.61	0.16	0.31	0.88	0.8
18:00	4 388	230	125	73	68	28.1	32.0	168.8	105	32.6	0.96	1.61	0.16	0.31	0.88	0.8
19:00	4 056	230	125	73	68	28.1	32.0	168.8	105	32.6	0.96	1.61	0.16	0.31	0.88	0.8
20:00	4 056	230	125	73	68	28.1	32.0	168.8	105	32.6	0.96	1.61	0.16	0.31	0.88	0.8
21:00	4 056	230	125	73	68	28.1	32.0	168.8	105	32.6	0.96	1.61	0.16	0.31	0.88	0.8
22:00	4 056	230	125	73	68	28.1	32.0	168.8	105	32.6	0.96	1.61	0.16	0.31	0.88	0.8
23:00	4 056	230	125	73	68	28.1	32.0	168.8	105	32.6	0.96	1.61	0.16	0.31	0.88	0.8
00:00	3 520	230	125	73	68	28.1	32.0	168.8	105	32.6	0.96	1.61	0.16	0.31	0.88	0.8
01:00	2 985	230	125	73	68	28.1	32.0	168.8	105	32.6	0.96	1.61	0.16	0.31	0.88	0.8
02:00	2 474	230	125	73	68	28.1	32.0	168.8	105	32.6	0.96	1.61	0.16	0.31	0.88	0.8
03:00	2 474	230	125	73	68	28.1	32.0	168.8	105	32.6	0.96	1.61	0.16	0.31	0.88	0.8
04:00	2 474	230	125	73	68	28.1	32.0	168.8	105	32.6	0.96	1.61	0.16	0.31	0.88	0.8
05:00	2 474	230	125	73	68	28.1	32.0	168.8	105	32.6	0.96	1.61	0.16	0.31	0.88	0.8
06:00	2 474	230	125	73	68	28.1	32.0	168.8	105	32.6	0.96	1.61	0.16	0.31	0.88	0.8

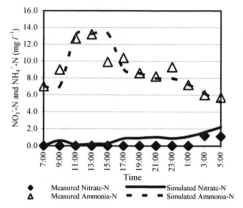

Figure 5.9(a). Experimental and simulated NH_4^+-N and NO_3^--N concentration profiles of the anoxic reactor in the secondary calibration under dynamic state conditions.

Figure 5.9(b). Experimental and simulated NH_4^+-N and NO_3^--N concentration profiles of the first aerobic reactor in the secondary calibration under dynamic state conditions.

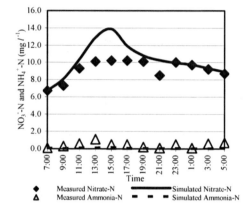

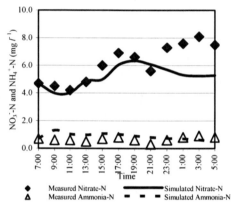

Figure 5.9(c). Experimental and simulated NH_4^+-N and NO_3^--N concentration profiles of the second aerobic reactor in the secondary calibration under dynamic state conditions.

Figure 5.9(d). Experimental and simulated NH_4^+-N and NO_3^--N concentration profiles of the RAS in the secondary calibration under dynamic state conditions.

Figures 5.9 (a)–(d) show the measured and simulated NH_4^+-N and NO_3^--N concentration profiles in the anoxic reactor, first aerobic reactor, second aerobic reactor and RAS, respectively. Variations in the simulated NO_3^--N profile in the anoxic reactor (Figure 5.9(a)) were satisfactorily consistent with the measured profile. The simulated NH_4^+-N profile approximated to the measured profile in the anoxic reactor although slight deviations were recorded between 11:00 and 16:00. These results indicated that the calibrated kinetic constants and stoichiometric coefficients of heterotrophs were suitable to describe denitrification under dynamic feed conditions. Figure 5.9(b) illustrates a general agreement between the simulated and measured profiles of NH_4^+-N concentrations but a degree of parity occurred between 11:00 and 16:00 in the first aerobic reactor. The discrepancy might have resulted from the lower

NH_4^+-N concentration derived during the same period from the preceding anoxic reactor. The higher simulated NO_3^--N concentration as compared to the measured concentrations could be observed in both the first and second aerobic reactors (Figures 5.9(b) and 5.9(c)). This was attributed to the lower measured NH_4^+-N concentration as compared to the simulated concentrations between 11:00 and 16:00 in the anoxic reactor. The measured NO_3^--N and NH_4^+-N concentration profiles approximated to the simulated concentration profiles in the RAS (Figure 5.9(d)) after b_H was reduced from 1.62 to 0.62 d^{-1}.

Table 5.9 shows the comparisons between the measured and simulated average NH_4^+-N and NO_3^--N concentrations in the three reactors. The differences between the measured and simulated average concentrations ranged between 0.2 and 1.1 mg NO_3^--N l^{-1}. The deviations in the NO_3^--N concentrations during the peak flow period were slightly high. Overall, the calibrated kinetic constants and stoichiometric coefficients of autotrophic nitrifiers obtained under steady-state feed conditions were able to describe nitrification under dynamic feed conditions, and no further calibration was made.

Table 5.9. Measured and calibrated average parameter concentrations (mg l^{-1}) and their differences in the reactors of the laboratory-scale MLE configuration under dynamic state conditions.

Value	NO_3^--N			NH_4^+-N		
	ANO	AER (1)	AER (2)	ANO	AER (1)	AER (2)
Measured	0.2	8.1	9.2	8.9	0.6	0.4
Calibrated	0.8	8.9	10.2	8.8	0.8	0.1
Difference	-0.6	-0.8	-1.0	0.1	-0.2	0.3

5.4   PARAMETER VERIFICATION OF ASM NO. 1

To explore the applicability and reliability of the values of the parameters calibrated from the laboratory experiment simulating the performance of the activated sludge process of one particular WRP (Bedok WRP in this case), the values of the calibrated parameters were verified extensively with three laboratory-scale and two full-scale measured diurnal data sets of the performances of the BNR activated sludge processes of the three WRPs investigated. The results are promising and encouraging (Cao et al., 2004c). Parts of the results are presented in the following sequence:

i   verification with the performance data of the laboratory-scale MLE process proposed for Battery B of Seletar WRP;

ii   verification with the performance data of the full-scale MLE process of Phase IV of Bedok WRP; and

iii   verification with the performance data of the full-scale MLE process of Phase III of Kranji WRP.

5.4.1 Verification of laboratory-scale modified MLE process of Battery B of Seletar WRP

The modified MLE process of the activated sludge process of Battery B of Selater WRP has 62.5% anoxic volume and 37.5% aerobic volume as shown in Figure 5.10. According to the detailed site study conducted (Cao, et al., 2004a), the diurnal hydraulic flow had two peaks, one during the day and the other at night, and a low (base) flow between 01:00 and 05:00. A peak flow factor of 1.9 was recorded (Cao, et al. 2004c). The calibrated kinetic constants and stoichiometric coefficients (Table 5.2) were used to describe BNR in the reactors and the CSTR used to simulate denitrification in the sludge blanket. The COD-based influent data file, including the diurnal feed flow, the compositions and conversion factors, was designed based on those data recorded at Battery B of Seletar WRP (Cao *et al.* 2004c).

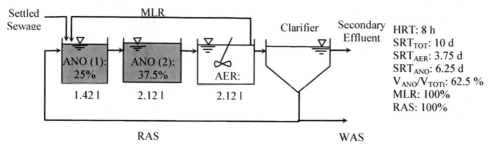

Figure 5.10. Schematic diagram of the laboratory-scale modified MLE process of Battery B of Seletar WRP.

Figures 5.11(a)–(d) show that the simulated NO_3^--N concentration profiles of the first and second anoxic reactors fitted satisfactorily with the measured concentration profiles, and indicated that the calibrated kinetic constants and stoichiometric coefficients of the heterotrophs were applicable. Figure 5.11(c) shows that the simulated NH_4^+-N concentration profile in the last aerobic reactor closely followed the trend of the measured concentration profile and showed that the kinetic constants and stoichiometric coefficients of the autotrophic nitrifier were applicable.

As shown in Table 5.10, which compiles the average values of the measured and simulated concentrations, the differences between the simulated and measured data were < 5%, illustrating that the kinetic constants and stoichiometric coefficients calibrated with the activated sludge process of Bedok WRP could be successfully applied to describe the activated sludge process in Seletar WRP.

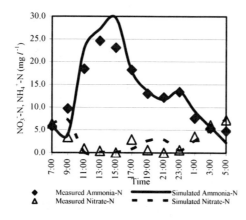

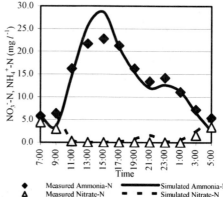

Figure 5.11(a). Experimental and simulated NH_4^+-N and NO_3^--N concentration profiles of the first anoxic reactor of the laboratory-scale modified MLE process of Battery B of Seletar WRP.

Figure 5.11(b). Experimental and simulated NH_4^+-N and NO_3^--N concentration profiles of the second anoxic reactor of the laboratory-scale modified MLE process of Battery B of Seletar WRP.

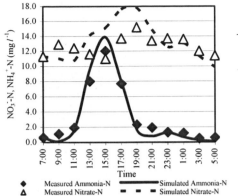
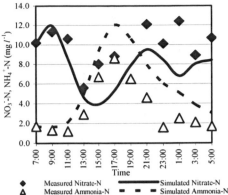

Figure 5.11(c). Experimental and simulated NH_4^+-N and NO_3^--N concentration profiles of the aerobic reactor of the laboratory-scale modified MLE process of Battery B of Seletar WRP.

Figure 5.11(d). Experimental and simulated NH_4^+-N and NO_3^--N concentration profiles of the RAS of the laboratory-scale modified MLE prcoess of Battery B of Seletar WRP.

Table 5.10. Measured and simulated average parameter concentrations (mg l^{-1}) and their differences in the reactors of the laboratory-scale modified MLE process of Battery B of Seletar WRP.

Value	NO_3^--N			NH_4^+-N		
	ANO (1)	ANO (2)	AER	ANO (1)	ANO (2)	AER
Measured	2.7	1.1	12.7	13.0	13.5	3.3
Simulated	3.0	1.8	13.3	13.6	13.3	3.2
Difference	-0.3	-0.7	-0.6	-0.6	0.2	0.1

5.4.2 Verification of full-scale MLE Process of Phase IV of Bedok WRP

From the point of view of scale-up, the real challenge is to verify the parameters calibrated using data of the laboratory-scale system(s) with those of the full-scale system(s). In this study, this was carried out by using the measured data from the two full-scale activated sludge processes of Bedok and Kranji WRPs. Chapter 3 details the investigation of the MLE activated sludge process of Bedok WRP, including the process configuration and key parameters (Figure 3.1), feed conditions (Section 3.3.2) and the measured data in individual compartments (Sections 3.3.2 to 3.3.5), and all were adopted in verification.

The compositions and conversion factors in the COD-based influent model, calculated according to the sewage characterization data and Table 5.4, are shown in Table 5.11. The kinetic constants and stoichiometric coefficients of the anoxic and aerobic compartments and the sludge blanket CSTR are those as listed in Table 5.2. Figures 5.12(a)–(e) show the measured and simulated concentration profiles of NH_4^+-N and NO_3^--N in the four compartments and the RAS.

Table 5.11. Diurnal influent data of the full-scale MLE process of Phase IV of Bedok WRP.

Time	q (m³ d⁻¹)	COD (mg l⁻¹)	SCOD (mg l⁻¹)	BOD₅ (mg l⁻¹)	TSS (mg l⁻¹)	NH₄⁺-N (mg l⁻¹)	TKN (mg l⁻¹)	ALK (mg l⁻¹)	Sₛ (mg l⁻¹)	Xₛ (mg l⁻¹)	ivt	icv	frsi	frxs	fnh	fxn
07:00	3 553	251	99	-	69.0	25.6	31.8	168.4	79	50	0.96	2.29	0.20	0.33	0.81	0.80
08:00	3 682	254	106	-	66.9	30.5	35.8	177.6	84	48	0.96	2.31	0.20	0.32	0.85	0.80
09:00	3 811	257	112	97	64.9	35.4	39.7	186.9	90	46	0.96	2.32	0.20	0.32	0.89	0.80
10:00	3 781	243	110	-	68.9	33.6	37.7	194.4	88	42	0.96	2.01	0.20	0.31	0.89	0.80
11:00	3 750	228	107	-	72.8	31.8	35.7	202.0	86	37	0.96	1.73	0.20	0.31	0.89	0.80
12:00	3 749	248	110	106	76.8	30.1	34.5	186.1	88	44	0.96	1.87	0.20	0.32	0.87	0.80
13:00	3 747	268	113	-	80.7	28.3	33.3	170.1	90	50	0.96	2.00	0.20	0.32	0.85	0.80
14:00	3 574	266	112	-	81.5	26.6	30.5	173.1	89	49	0.96	1.96	0.20	0.32	0.87	0.80
15:00	3 400	263	110	-	82.4	24.8	27.7	176.0	88	49	0.96	1.93	0.20	0.32	0.90	0.80
16:00	3 421	277	118	-	83.2	23.3	29.5	170.1	94	51	0.96	1.99	0.20	0.32	0.79	0.80
17:00	3 442	291	126	104	84.0	21.7	31.2	164.3	101	52	0.96	2.04	0.20	0.32	0.70	0.80
18:00	3 521	293	120	-	86.1	22.5	31.7	165.5	96	56	0.96	2.10	0.20	0.32	0.71	0.80
19:00	3 600	295	113	-	88.2	23.2	32.1	166.8	90	60	0.96	2.15	0.20	0.33	0.72	0.80
20:00	3 676	286	108	-	90.2	22.7	34.7	165.3	86	59	0.96	2.06	0.20	0.33	0.65	0.80
21:00	3 753	277	102	98	92.3	22.1	37.2	163.8	82	58	0.96	1.97	0.20	0.33	0.59	0.80
22:00	3 560	267	103	-	86.2	22.2	34.1	161.1	82	54	0.96	1.98	0.20	0.33	0.65	0.80
23:00	3 367	256	103	-	80.0	22.2	31.0	158.4	82	50	0.96	1.99	0.20	0.32	0.72	0.80
00:00	3 139	261	111	-	73.9	22.4	32.3	160.9	88	48	0.96	2.12	0.20	0.32	0.69	0.80
01:00	2 911	266	118	104	67.7	22.6	33.6	163.4	94	47	0.96	2.27	0.20	0.32	0.67	0.80
02:00	2 342	264	112	-	69.0	22.9	32.5	167.8	90	49	0.96	2.29	0.20	0.32	0.70	0.80
03:00	1 772	263	106	-	70.4	23.2	31.5	172.2	85	51	0.96	2.31	0.20	0.32	0.74	0.80
04:00	1 910	261	100	-	71.7	23.4	30.4	176.6	80	53	0.96	2.33	0.20	0.33	0.77	0.80
05:00	2 047	259	94	87	73.0	23.7	29.3	181.0	75	55	0.96	2.35	0.20	0.33	0.81	0.80
06:00	2 800	255	97	-	71.0	24.7	30.6	174.7	77	52	0.96	2.32	0.20	0.33	0.81	0.80

q: Flow rate to a 1 000 m³ tank
frxs: Degradable fraction of particulate COD
ivt: VSS to TSS ratio
fnh: NH₄⁺-N to TKN ratio
icv: Particulate COD to VSS ratio
fxn: Particulate organic nitrogen to total organic nitrogen ratio
frsi: Inert fraction of soluble COD

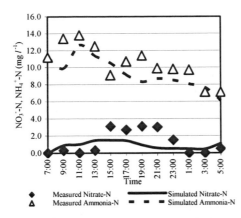 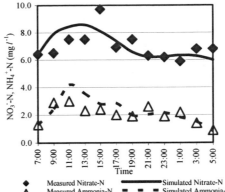

Figure 5.12(a). Experimental and simulated NH₄⁺-N and NO₃⁻-N concentration profiles of the anoxic zone of Phase IV of Bedok WRP.

Figure 5.12(b). Experimental and simulated NH₄⁺-N and NO₃⁻-N concentration profiles of the first aerobic zone of Phase IV of Bedok WRP.

As shown in Figure 5.12(a), the general tendency of the measured and simulated NH_4^+-N and NO_3^--N concentration profiles in the anoxic compartment fitted each other satisfactorily. The simulated NO_3^--N concentrations were 1.5-2.0 mg NO_3^--N l^{-1} lower than the measured values between 15:00 and 23:00, and the simulated NH_4^+-N concentrations were 1.0-2.0 mg NH_4^+-N l^{-1} lower than the measured values between 17:00 and 01:00. These deviations might have resulted due to the load transient discrepancies between the influent, the model and the MLR and RAS (Langeveld, 2004). Influent flows into individual trains were calculated assuming uniform distribution among the 8 trains since flow meters in the individual trains were not available. However, in reality, the actual flow into each train might not have been equal and the pump readings used for the MLR and RAS might not have been accurate. Despite these possible variations in the operation data, verification was still satisfactory and substantiated the applicability of the kinetic constants and stoichiometric coefficients of the heterotrophic bacteria to describe the full-scale system.

The simulated NH_4^+-N and NO_3^--N concentration profiles in the first aerobic compartment agreed well with the measured concentration profiles (Figure 5.12(b)). This indicated that the kinetic constants and stoichiometric coefficients of the autotrophic nitrifiers could predict nitrification in the full-scale activated sludge process.

The measured and simulated NH_4^+-N and NO_3^--N concentration profiles in the second and third aerobic compartments fitted satisfactorily (Figures 5.12(c) and (d)). The NH_4^+-N concentration reduced significantly from the anoxic to the aerobic reactors. The simulation indicated that the NH_4^+-N concentration was < 1 mg NH_4^+-N l^{-1} for most of the time and no marked reduction in the third aerobic compartment was apparent, illustrating the possibility of omitting the last compartment. Also, the simulated NH_4^+-N concentrations were lower than the measured values in the last two aerobic compartments, most likely, due to inhibition of the lower pH (ranged between 6.0 and 6.4) of the liquor in the aerated compartments resulting from the lower alkalinity of the settled sewage (Table 2.1; Section 2.3.2), which ASM No. 1 does not account for.

Figure 5.12(e) shows the measured and simulated NH_4^+-N and NO_3^--N concentration profiles in the RAS. The deviations between the simulated and the measured concentration profiles were larger than those recorded in the verifications with the laboratory-scale data. In addition to the simplified models adopted, another reason for the deviations could have been that the secondary clarifier received MLSS from two treatment trains, in which one train employed MLR and the other did not.

The differences between the measured and simulated average concentrations are presented in Table 5.12. The range differences were comparable with those reported by Concha and Henze (1992 and 1996), who measured and simulated data of the same pilot-scale activated sludge facility, while this simulation study was performed using data from a 6-litre reactor system which was scaled-down from an activated sludge process with a volume in the range of 8 700 m^3.

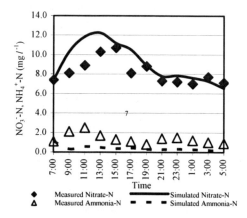

Figure 5.12(c). Experimental and simulated NH_4^+-N and NO_3^--N concentration profiles of the second aerobic zone of Phase IV of Bedok WRP.

Figure 5.12(d). Experimental and simulated NH_4^+-N and NO_3^--N concentration profiles of the third aerobic zone of Phase IV of Bedok WRP.

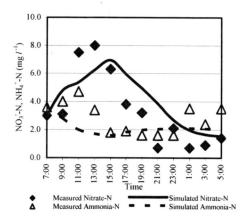

Figure 5.12(e). Experimental and simulated NH_4^+-N and NO_3^--N concentration profiles of the RAS of Phase IV of Bedok WRP.

Table 5.12. Measured and simulated average parameter concentrations (mg l⁻¹) and their differences in the anoxic and three aerobic zones of Phase IV of Bedok WRP.

Value	NO₃⁻-N				NH₄⁺-N			
	ANO	AER(1)	AER(2)	AER(3)	ANO	AER(1)	AER(2)	AER(3)
Measured	1.2	7.0	8.2	8.9	10.5	2.1	1.4	1.6
Simulated	0.9	7.0	9.1	9.7	9.3	2.3	0.4	0.2
Difference	0.3	0.0	1.1	-0.8	1.2	-0.2	1.0	1.4

5.4.3 Verification of full-scale MLE process of Phase III of Kranji WRP

The activated sludge process of Phase III of Kranji WRP is a MLE process. The effective volume of activated sludge tanks of a single train is 6 150 m³ with dimensions of 56 m x 11 m x 10 m (length x width x depth). It is partitioned into seven compartments. The anoxic zone accounts 20% of the total volume. The key parameters include HRT = 12 h, temperature = 30±1⁰, SRT_{TOT} = 10 d and MLSS = 2 500-3 000 mg l⁻¹. Air diffusers are used for aeration. Simultaneous nitrification and denitrification was recorded in the aerobic compartments in which the DO concentration varied between 0.8 and 1.8 mg l⁻¹. The diurnal hydraulic flow recorded two peaks, one during the day and the other at night. The low (base) flow occurred from 03:00 and 07:00. Figure 5.13 shows the schematic diagram and the designed operation conditions of the process.

A detailed site investigation was conducted in November 2002 (Cao *et al.* 2003). The measured data were used for full-scale verification. The process configuration, as shown in Figure 5.13, was adopted in modeling. The diurnal hydraulic flow, influent sewage composition, initialization values and other conditions were used to formulate a COD-based influent model according to the sewage characterization data and those of Table 5.4, are shown in Table 5.13. The kinetic constants and stoichiometric coefficients of the models representing the anoxic and aerobic compartments and the sludge blanket are listed in Table 5.2 with the exception that the b_{II} value of the sludge blanket CSTR, 0.72 d⁻¹, was adopted for better fitting.

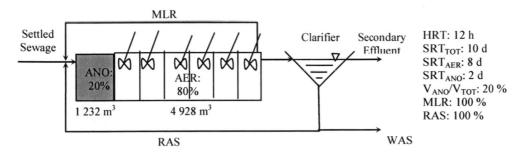

Figure 5.13. Schematic diagram of the full-scale MLE process of Phase III of Kranji WRP.

Table 5.13. Diurnal influent data of the full-scale MLE process of Phase III of Kranji WRP.

Time	q (m³ d⁻¹)	COD (mg l⁻¹)	SCOD (mg l⁻¹)	BOD₅ (mg l⁻¹)	TSS (mg l⁻¹)	NH₄⁺-N (mg l⁻¹)	TKN (mg l⁻¹)	ALK (mg l⁻¹)	S_S (mg l⁻¹)	X_S (mg l⁻¹)	ivt	icv	frsi	frxs	fnh	fxn
07:00	662	375	135	162	141.0	27.7	42.6	172.5	116	64	0.96	1.77	0.14	0.27	0.65	0.80
08:00	1 857	330	125	-	141.5	32.5	44.4	182.9	108	51	0.96	1.51	0.14	0.25	0.73	0.80
09:00	2 671	345	133	121	142.0	40.2	50.8	204.8	114	51	0.96	1.55	0.14	0.24	0.79	0.80
10:00	3 345	354	145	-	145.7	41.4	53.8	209.8	125	45	0.96	1.49	0.14	0.22	0.77	0.80
11:00	3 799	622	160	233	302.0	46.2	65.0	226.7	138	161	0.96	1.59	0.14	0.35	0.71	0.80
12:00	3 867	680	161	-	280.5	40.0	59.4	205.3	138	188	0.96	1.92	0.14	0.36	0.67	0.80
13:00	3 339	661	166	159	266.2	36.4	55.5	198.7	142	175	0.96	1.94	0.14	0.35	0.66	0.80
14:00	2 952	607	165	226	254.7	32.9	51.0	189.0	142	149	0.96	1.80	0.14	0.34	0.64	0.80
15:00	2 705	516	159	201	246.0	29.4	45.9	175.9	137	111	0.96	1.51	0.14	0.31	0.64	0.80
16:00	2 308	563	156	-	248.0	33.4	47.1	180.5	134	136	0.96	1.71	0.14	0.33	0.71	0.80
17:00	2 671	582	176	189	250.0	33.6	47.3	181.8	151	128	0.96	1.69	0.14	0.32	0.71	0.80
18:00	2 386	509	164	-	227.5	27.3	43.5	162.0	141	103	0.96	1.58	0.14	0.30	0.63	0.80
19:00	2 197	439	154	149	205.0	27.9	43.9	168.8	132	78	0.96	1.45	0.14	0.27	0.64	0.80
20:00	2 853	428	157	-	199.9	26.4	40.9	160.4	135	70	0.96	1.41	0.14	0.26	0.65	0.80
21:00	3 571	783	160	247	371.0	24.1	49.3	155.7	138	238	0.96	1.75	0.14	0.38	0.49	0.80
22:00	3 808	880	163	-	304.0	27.1	55.0	172.6	140	282	0.96	2.45	0.14	0.39	0.49	0.80
23:00	3 866	570	144	207	237.0	25.5	42.1	157.2	124	150	0.96	1.87	0.14	0.35	0.61	0.80
00:00	3 563	406	139	-	184.0	23.0	38.4	146.8	120	75	0.96	1.51	0.14	0.28	0.60	0.80
01:00	3 301	347	129	128	131.0	24.0	34.6	152.5	111	56	0.96	1.73	0.14	0.26	0.69	0.80
02:00	2 733	343	127	-	138.0	28.9	39.4	172.8	109	55	0.96	1.63	0.14	0.26	0.73	0.80
03:00	1 649	345	125	134	145.0	28.2	40.0	175.4	108	58	0.96	1.58	0.14	0.26	0.71	0.80
04:00	849	342	132	-	145.0	28.8	40.8	171.8	114	51	0.96	1.51	0.14	0.24	0.71	0.80
05:00	592	348	156	122	145.0	26.3	37.9	161.2	134	33	0.96	1.38	0.14	0.17	0.69	0.80
06:00	503	438	142	-	143.0	26.4	43.9	162.4	122	88	0.96	2.15	0.14	0.30	0.60	0.80

q: Flow rate to a 1 000 m³ tank
ivt: VSS to TSS ratio
icv: Particulate COD to VSS ratio
frsi: Inert fraction of soluble COD
frxs: Degradable fraction of particulate COD
fnh: NH₄⁺-N to TKN ratio
fxn: Particulate organic nitrogen to total organic nitrogen ratio

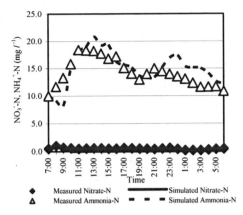

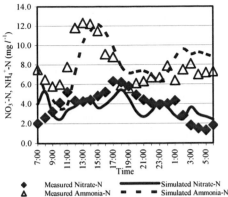

Figure 5.14(a). Experimental and simulated NH$_4^+$-N and NO$_3^-$-N concentration profiles of the anoxic zone of Phase III of Kranji WRP.

Figure 5.14(b). Experimental and simulated NH$_4^+$-N and NO$_3^-$-N concentration profiles of the last aerobic zone of Phase III of Kranji WRP.

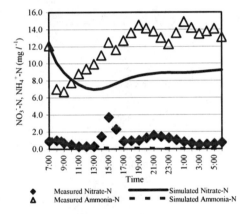

Figure 5.14(c). Experimental and simulated NH$_4^+$-N and NO$_3^-$-N concentration profiles of the RAS of Phase III of Kranji WRP.

Figures 5.14(a) – (c) show the measured and simulated NH$_4^+$-N and NO$_3^-$-N concentration profiles in the anoxic compartment, the last aerobic compartment and the RAS, respectively.

As shown in Figure 5.14(a), the simulated NH$_4^+$-N and NO$_3^-$-N concentration profiles in the anoxic compartment fitted reasonably well with the measured concentration profiles. Also, with the exception of between 10:00 and 12:00, the simulated NH$_4^+$-N and NO$_3^-$-N concentration profiles in the last aerobic compartment agreed well with the measured concentration profiles. The fluctuations of the DO concentration between 0.5 and 1.8 mg O$_2$ l^{-1} in the aerobic compartment, which greatly affected simultaneous nitrification and denitrification, could have been the main cause. The reasons for the deviations between the simulated and measured concentration profiles in the RAS were the results of the simplified model adopted for the secondary clarifier as mentioned previously.

Table 5.14 compiles the measured and simulated average parameter concentrations and their differences, which were in the acceptable range and, hence, verification was deemed successful.

Table 5.14. Measured and simulated average parameter concentrations (mg l^{-1}) and their differences in the anoxic zone and last aerobic zone of Phase III of Kranji WRP.

Value	NO_3^--N		NH_4^+-N	
	ANO	AER (6)	ANO	AER (6)
Measured	0.5	3.9	14.2	7.8
Simulated	0.1	3.6	15.0	8.0
Difference	0.4	0.3	-0.8	-0.2

The results of verification indicated that the parameters calibrated by the laboratory experiments were able to describe BNR in the full-scale activated sludge processes of the two water reclamation plants. Also, it illustrated that both the scale-down design of the laboratory experiment and the application of the COD-based influent model under Singapore conditions were successful.

5.5 SIMULATION OF BNR ACTIVATED SLUDGE PROCESS BY VERIFIED ASM NO. 1

The ASM No. 1 with verified parameters and coefficients was used for investigations on several issues of the BNR activated sludge process that were not easy to study by experimentation, such as effects of DO concentration, COD/TKN ratio and fraction of the anoxic volume on denitrification in anoxic zone, spatial distribution of heterotrophic bacteria and autotrophic nitrifiers in the activated sludge process.

The MLE activated sludge process configuration of Phase IV of Bedok WRP and the typical feed and operation conditions under steady state were adopted for most of the simulations. Apparently, the outcomes and the discussions of the simulations have a wider scope of application on operation and design and were not necessarily limited to the MLE process in a warm climate.

5.5.1 Critical issues in the BNR activated sludge processes

5.5.1.1 Effect of DO concentration in anoxic zone

The DO concentration has a direct effect on denitrification in an anoxic zone since heterotrophic bacteria prefer to use oxygen as the electron acceptor compared with NO_3^--N. A simulation was made to quantify this effect with the MLE activated sludge process of Phase IV of Bedok WRP fed with a constant flow and composition as detailed in Cao et al., (2004c). As shown in Figure 5.15, the influent-based NO_3^--N removal in the anoxic zone decreased from 23.4 to 1.1 mg NO_3^--N l^{-1} when the DO concentration in the anoxic compartment increased from 0 to 0.12 mg O_2 l^{-1}. It should be noted that the DO concentration quoted here referred to the concentration in the whole anoxic compartment while the measured DO concentration that ranged between 0.1 and 0.3 mg O_2 l^{-1} was the value near the surface, which was higher than that used in the simulation. However, this exemplified the necessity to control the DO concentration in the last compartment, the importance of the location of MLR pipe outlet into the anoxic zone and reduction of disturbances at the surface of the anoxic compartment.

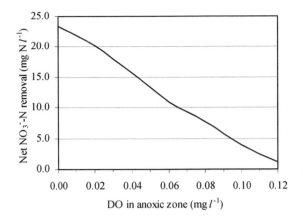

Figure 5.15. Influent-based NO$_3^-$-N removal in the anoxic zone when DO concentration increased from 0 to 0.12 mg l^{-1}.

5.5.1.2 Effect of MLR ratio

The MLR influences denitrification in an anoxic zone since it offers another source of NO$_3^-$-N in addition to that from the RAS. To study the effect of MLR ratio, a simulation was made with the MLE activated sludge process of Phase IV of Bedok WRP, which was fed with the following typical feed composition: COD, 294 mg l^{-1}; SCOD, 114 mg l^{-1} and other parameters and coefficients of the COD-based influent model (Cao *et al.*, 2004c). The DO concentration in the anoxic zone varied with the MLR ratio but remained < 0.02 mg O$_2$ l^{-1} in the simulated range. Figure 5.16 shows the relationship between the MLR ratio and the influent-based NO$_3^-$-N removal concentration in the anoxic compartment.

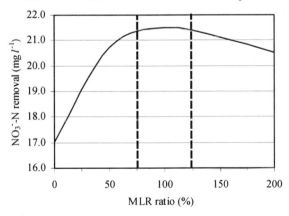

Figure 5.16. Influent-based NO$_3^-$-N removal when the MLR ratio increased from 0 to 200% in the anoxic zone.

Three regions can be distinguished according to the effects of the MLR ratio on denitrification as shown in Figure 5.16. For a MLR ratio < 75%, NO$_3^-$-N removal increased as the MLR ratio increased since NO$_3^-$-N removal was nitrate-limited in this region; there was a benefit to introducing MLR operation. For a MLR ratio > 125%, NO$_3^-$-N removal did not increase but reduced. Denitrification was then carbon-limited. The reduction in denitrification was most likely caused by the low substrate due to dilution and increased DO concentrations. For a MLR ratio varying between 75 and 125%, a maximum of 21.5 mg NO_3^--N l^{-1} was removed in this dual-limitation region. These phenomena were consistent with those of the laboratory experiment (Cao, *et al.* 2004a). Therefore, a 100% MLR ratio was recommended to operate the MLE process under Singapore conditions.

5.5.1.3 Effect of COD/TKN ratio

The influent sewage COD/TKN ratio is regarded as an indicator of denitrification potential (Metcalf & Eddy, 2003). In fact, in addition to the COD, the biodegradable COD fractions (S_S and X_S) should be considered when estimating the denitrification potential. Simulations were made with the MLE activated sludge process of Bedok WRP with a 25% anoxic volume and MLR ratios between 50 and 200% to study the effects of the COD/TKN ratio. In the simulations, the ratio was varied from 5 to 10 by increasing the COD and maintaining a TKN concentration of 50 mg N l^{-1}. When the COD was increased from 250 to 500 mg COD l^{-1}, the S_S increased from 80 to 180 mg COD l^{-1} and the X_S from 45 to 70 mg COD l^{-1}, and some conversion coefficients changed as well (Cao *et al.*, 2004c). The influent flow and composition used were typical values of the influent sewage of Bedok WRP (Section 2.3.2). In the simulations, the DO concentration in the aerobic zone was controlled at 2.0 mg O_2 l^{-1} while that of the anoxic zone was < 0.01 mg O_2 l^{-1}.

As shown in Figure 5.17, the whole process can be divided into three regimes: (i) carbon-limited; (ii) nitrate-limited; and (iii) dual-limitation. For each MLR ratio considered, the COD/TKN ratios to the left of the point of maximum NO_3^--N removal were in the carbon-limited regime while those to the right were in the nitrate- limited regime. The higher the MLR ratio, the higher was the COD/TKN ratio corresponding to maximum NO_3^--N removal. For the MLR ratios considered, the NO_3^--N removal decreased with the COD/TKN ratio increase after the maximum removal was reached. This might have resulted due to more nitrogen assimilation into biomass and a subsequent reduction of NO_3^--N formation from TKN hydrolysis and oxidation.

The simulations showed that under carbon-limited conditions (at low COD/TKN ratios), a lower MLR ratio effected higher NO_3^--N removals while a higher MLR ratio gave lower NO_3^--N removals (Figure 5.17). This was due to the increased DO concentration effect along with a higher MLR ratio on denitrification that was discussed previously. However, under nitrate-limited conditions (at high COD/TKN ratios), the higher the MLR ratio, the higher the NO_3^--N removal due to the fact that with higher COD/TKN ratios, high S_S (mainly) and X_S were available to facilitate the denitrification of increased concentrations of NO_3^--N recycled by the high MLR ratio. The effluent TN concentrations decreased when the NO_3^--N removals increased in the anoxic zone.

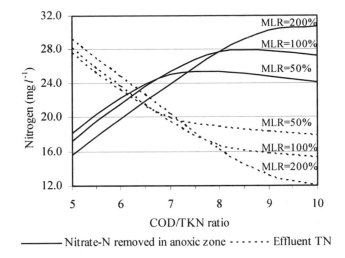

Figure 5.17. NO$_3^-$-N removals in the anoxic zone and the effluent TN in response to varying influent COD/TKN ratios and MLR ratios of 50, 100 and 200%.

5.5.1.4 Effect of anoxic volumetric ratio

The volume of anoxic zone is related to the anoxic sludge retention time (SRT$_{ANO}$), one of the determining factors of NO$_3^-$-N removal. Simulations were made with ASM No. 1 with verified parameter values to identify the optimal anoxic volume under the conditions of a total SRT of 10 d, a HRT of 6 h, and RAS and MLR ratios of 100% each, with three influent compositions comprising different combinations of S$_S$, X$_S$ and other parameters and coefficients of the COD-based influent model.

As shown in Figure 5.18, three regions were distinguished according to the changes in denitrification in response to anoxic volume variations. When the anoxic volume fraction was increased from the site ratio of 25% to 50%, with an influent composition of a S$_S$ of 110 mg COD l^{-1} and a X$_S$ of 52.5 mg COD l^{-1}, the NO$_3^-$-N concentration of the anoxic compartment effluent fell from 7.5 to 1.3 mg NO$_3^-$-N l^{-1} while the influent-based NO$_3^-$-N removal increased from 22.0 to 26.5 mg NO$_3^-$-N l^{-1}. For the influent composition with a Ss of 70 mg COD l^{-1} and a X$_S$ of 92.5 mg COD l^{-1}, the effluent NO$_3^-$-N concentration fell from 10.5 to 2.3 mg NO$_3^-$-N l^{-1} while the anoxic NO$_3^-$-N removal increased from 19.0 to 23.5 mg NO$_3^-$-N l^{-1}. Finally, for the influent composition with a S$_S$ of 30 mg COD l^{-1} and a X$_S$ of 132.5 mg COD l^{-1}, the effluent NO$_3^-$-N concentration fell from 10.6 to 0.8 mg NO$_3^-$-N l^{-1} while the anoxic NO$_3^-$-N removal increased from 18.5 to 26.5 mg NO$_3^-$-N l^{-1}.

The influent sewage composition with the highest S$_S$ resulted in the highest removal of NO$_3^-$-N. However, the highest increase in denitrification due to the anoxic volume fraction increase coincided with the highest X$_S$. Considering the above results, it could be concluded that the increase of NO$_3^-$-N removal with the anoxic volume increase was mainly due to the utilization of biodegradable particulate COD (X$_S$). This was because S$_S$ was utilized within 30 minutes (EPA, 1993) while it took about 3 to 5 times longer for the hydrolysis and utilization of particulate COD (X$_S$) for denitrification (Table 2.4).

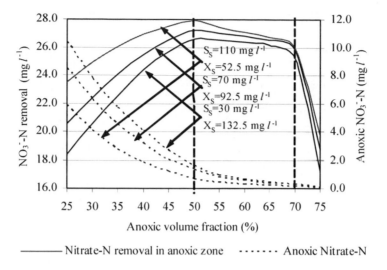

Figure 5.18. Relationship between NO₃⁻-N removed and the NO₃⁻-N concentration in the anoxic compartment with the anoxic volume fraction.

When the anoxic volume fraction varied between 50 and 70%, the influent-based NO₃⁻-N removal concentration recorded a decline while the anoxic reactor effluent NO₃⁻-N concentration continued to fall. This phenomenon was probably related to the slight decline in ammonia oxidation due to the decline in autotrophic nitrifiers in the aerobic zone as a result of the decrease of the aerobic SRT. When the anoxic volume fraction was > 70%, a sharp decline in denitrification was observed due to the triggering of the 'wash out' of autotrophic nitrifier, which was induced by an insufficient aerobic SRT. Therefore, an anoxic volume fraction of 50% corresponding to a 5 d anoxic SRT is recommended for the MLE processes under Singapore conditions. When the anoxic volumetric ratio increased from 25 to 50%, on average, the NO₃⁻-N concentration in the anoxic compartment reduced by 7.7 mg NO₃⁻-N l⁻¹ while the influent-based NO₃⁻-N removal concentration increased by 5.8 mg NO₃⁻-N l⁻¹, which corresponded to a 25% increase in denitrification. This was consistent with the experimental results (Cao *et al.* 2004a). A further analysis of NO₃⁻-N removal with increased anoxic volume and possible energy savings in the improved MLE process will be discussed in Section 5.5.2.

5.5.1.5 Spatial distribution of autotrophic nitrifiers

A simulation was made with the MLE activated sludge process of Bedok WRP, with the site operation conditions (Figure 3.1) and typical influent composition of COD of 325 mg l⁻¹, SCOD of 130 mg l⁻¹, TSS of 105 mg l⁻¹ and TKN of 50 mg N l⁻¹, to analyze the distribution of autotrophic nitrifiers. The simulated nitrifier concentrations in the anoxic and the three aerobic compartments were 149, 152, 152 and 151 mg COD l⁻¹, respectively (Figure 5.19). The variations were insignificant, even when the nitrifier concentration in the anoxic zone (with no nitrifier growth) was compared with those of the aerobic zones. These figures were consistent with the measured nitrification potentials of the sludge taken from the different zones of the activated sludge process of Bedok WRP as presented in Section 2.3.6. The slow growth rate of autotrophs compared with heterotrophs could have been the reason.

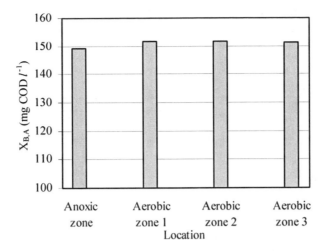

Figure 5.19. Simulated autotrophic nitrifier concentrations in the four zones of the activated sludge process of Phase IV of Bedok WRP.

The ASM No. 1 with verified parameters and coefficients was used to simulate the response of the nitrifier population to increases in the influent TKN load under steady-state feed conditions. As shown in Figure 5.20, the nitrifier concentrations in the reactor and the RAS increased linearly with the influent TKN concentration increases.

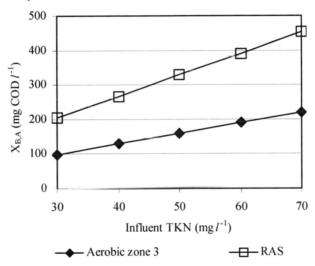

Figure 5.20. Variations of the autotrophic nitrifier concentrations with the influent TKN in the third aerobic zone and sludge blanket.

5.5.1.6 Minimum aerobic SRT for nitrification

The minimum aerobic volume or SRT for nitrification is governed by the kinetics of nitrification, which is temperature sensitive, and the operation conditions such as the aerobic dissolved oxygen concentration. Figure 5.21 shows the results of a simulation made to investigate the relationships between the aerobic SRT (related to the aerobic volume ratio) and the autotrophic nitrifier ($X_{B,A}$) concentration in the aerobic reactor as well as the effluent NH_4^+-N concentrations with the MLE process under the operation conditions in Figure 3.1 and the influent sewage composition of COD of 294 mg l^{-1}, SCOD of 114 mg l^{-1}, TSS of 109 mg l^{-1} and TKN of 47 mg N l^{-1}.

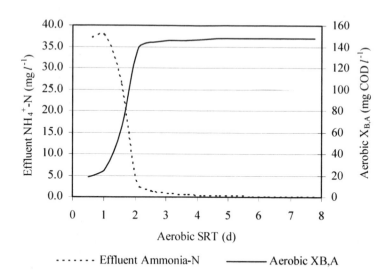

Figure 5.21. Variations of the effluent NH_4^+-N concentration and the third aerobic zone autotrophic nitrifier concentrations with the aerobic SRT.

Figure 5.21 indicates that the nitrifier concentration slightly declined, and concomitantly the effluent NH_4^+-N concentration increased when the aerobic SRT was reduced from 5 days. The nitrifiers were 'washed out' and the effluent NH_4^+-N concentration began to increase drastically when the aerobic SRT decreased to < 2.0 d. This value was close to the SRT value of 1.7 days calculated from the equation (Grady *et al.*, 1999):

$$(K_{NH} + NH_4^+\text{-}N_{EFF})/[NH_4^+\text{-}N_{EFF} \times (\hat{\mu}_A - b_A) - K_{NH} \times b_A]$$

where NH_4^+-N_{EFF} was assumed as 1 mg NH_4^+-N l^{-1} and the others were taken from Table 5.6. Thus, a 2 d SRT is regarded as the minimum aerobic SRT for nitrification in warm climates, and so the aerobic SRT should not be less than 2 d under Singapore conditions.

Considering the peak factor, an aerobic SRT of about 4 days for nitrification can be adopted in the design of the BNR activated sludge process in warm climates. An anoxic SRT between 3 and 5 days is proposed for denitrification. It is longer than that of conventional design (~ 20% of total volume), and is appropriate since both the readily biodegradable COD and the sewage alkalinity are near the lower boundary of the respective ranges of reported values under Singapore conditions. The final selection of anoxic SRT depends on the COD in the influent, the higher the COD, the shorter the anoxic SRT and vice versa. Table 5.15 compiles the key design parameters for the BNR activated sludge process in warm climates.

Table 5.15. Suggested design guidelines for the MLE activated sludge process in warm climates.

Parameter	Description	Unit	Value	Remark
Θ_{AER}	Aerobic sludge retention time (SRT_{AER})	d	4	$\Theta_{MIN\text{-}AER} = 2.0$ d, Corresponds to 50-60% of activated sludge tank volume.
Θ_{ANO}	Anoxic sludge retention time (SRT_{ANO})	d	3–5	Corresponds to 30 - 50% of activated sludge tank volume depending on COD in influent.
HRT	Hydraulic retention time	h	6–8	Based on the average influent flow and mass loads.
RAS	Return activated sludge	%	100	Based on the average influent flow. Smaller ratio can be adopted during base flow period.
MLR	Mixed liquor recycle	%	80–100	Based on the average influent flow. Smaller ratio can be adopted during base flow period.
Aeration	Oxygen for COD removal/nitrification	-	-	Maximum at the head of aerobic zone and gradual reduction towards the end.

5.5.2 Performance of the recommended upgrade of the MLE process

5.5.2.1 Process configuration

Based on the results of Section 5.5.1 and the studies of the existing activated sludge processes of Bedok, Kranji and Seletar WRPs and the laboratory experiment, a MLE activated sludge process configuration partitioned into four compartments of equal volume each, with anoxic and aerobic volumetric ratio of 50% each, SRT_{TOT} of 10 d, HRT of 6 h, and MLR and RAS ratio of 100% each, is proposed as the recommended configuration for the existing activated sludge process to be upgraded to. The schematic diagram of the process with the reactor sizes and operation conditions is given in Figure 5.22. Compared with the existing processes, the anoxic volumetric ratio was doubled, while the aerobic ratio was reduced by 50%, whereas both the total SRT and HRT were similar to the existing processes. This section investigates the performance of the recommended MLE configuration under dynamic (diurnal) load conditions. The DO concentration in each of the two anoxic compartments was set at 0.04 and 0.06 mg O_2 l^{-1}, respectively, and at 2 mg O_2 l^{-1} in each of the two aerobic compartments.

Simulations were made with peak factors between 1.2 and 2.0 with a fixed average HRT of 6 h (Figure 5.23). The influent composition and diurnal flow pattern were also similar under the different peak flow conditions. Table 5.16 shows the influent input data.

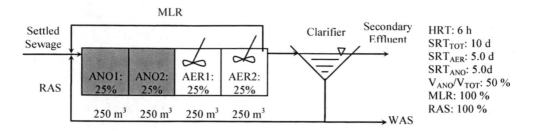

Figure 5.22. Schematic diagram of the recommended MLE configuration used in the simulation.

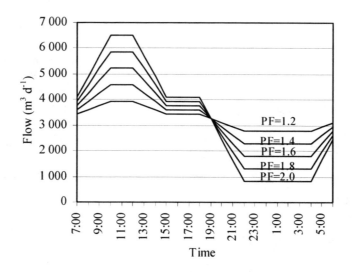

Figure 5.23. Diurnal hydraulic flow profiles with peak factors of 1.2, 1.4, 1.6, 1.8 and 2.0.

Table 5.16. Diurnal influent data used in the simulation of the MLE configuration of Bedok WRP.

Time	Sewage Flow (m³ d⁻¹)					icv gCOD (gVSS)⁻¹	COD g m⁻³	TSS g m⁻³	TKN g N m⁻³	ivt gVSS (gTSS)⁻¹	frsi	frxs	fnh	fxn	S$_{ALK}$ mole m⁻³	vmras m³	DO Set Point (mg l⁻¹)			
	PF=1.2	PF=1.4	PF=1.6	PF=1.8	PF=2.0												ANO (1)	ANO (2)	AER (1)	AER (2)
07:00	3 426	3 589	3 752	3 915	4 078	2.29	251.0	69.0	47.7	0.962	0.2	0.338	0.75	0.8	3.368	100	0.06	0.04	2.0	2.0
08:00	3 589	3 915	4 242	4 568	4 894	2.31	254.0	66.9	53.6	0.962	0.2	0.321	0.75	0.8	3.553	100	0.06	0.04	2.0	2.0
09:00	3 752	4 242	4 731	5 220	5 710	2.32	257.0	64.9	59.6	0.962	0.2	0.304	0.75	0.8	3.738	100	0.06	0.04	2.0	2.0
10:00	3 915	4 568	5 220	5 873	6 525	2.01	242.5	68.9	56.6	0.962	0.2	0.289	0.75	0.8	3.889	100	0.06	0.04	2.0	2.0
11:00	3 915	4 568	5 220	5 873	6 525	1.73	228.0	72.8	53.6	0.962	0.2	0.272	0.75	0.8	4.040	100	0.06	0.04	2.0	2.0
12:00	3 915	4 568	5 220	5 873	6 525	1.87	248.0	76.8	51.8	0.962	0.2	0.297	0.75	0.8	3.721	100	0.06	0.04	2.0	2.0
13:00	3 752	4 242	4 731	5 220	5 710	2.00	268.0	80.7	50.0	0.962	0.2	0.316	0.75	0.8	3.402	100	0.06	0.04	2.0	2.0
14:00	3 589	3 915	4 242	4 568	4 894	1.97	265.5	81.5	45.8	0.962	0.2	0.317	0.75	0.8	3.461	100	0.06	0.04	2.0	2.0
15:00	3 426	3 589	3 752	3 915	4 078	1.93	263.0	82.4	42.9	0.962	0.2	0.319	0.75	0.8	3.520	100	0.06	0.04	2.0	2.0
16:00	3 426	3 589	3 752	3 915	4 078	1.99	277.0	83.2	44.2	0.962	0.2	0.312	0.75	0.8	3.403	100	0.06	0.04	2.0	2.0
17:00	3 426	3 589	3 752	3 915	4 078	2.04	291.0	84.0	40.4	0.962	0.2	0.306	0.75	0.8	3.286	100	0.06	0.04	2.0	2.0
18:00	3 426	3 589	3 752	3 915	4 078	2.10	293.0	86.1	40.1	0.962	0.2	0.327	0.75	0.8	3.311	100	0.06	0.04	2.0	2.0
19:00	3 263	3 263	3 263	3 263	3 263	2.15	295.0	88.2	41.0	0.962	0.2	0.346	0.75	0.8	3.336	100	0.06	0.04	2.0	2.0
20:00	3 100	2 936	2 773	2 610	2 447	2.06	286.0	90.2	42.4	0.962	0.2	0.351	0.75	0.8	3.306	100	0.06	0.04	2.0	2.0
21:00	2 936	2 610	2 284	1 958	1 631	1.97	277.0	92.3	42.8	0.962	0.2	0.357	0.75	0.8	3.276	100	0.06	0.04	2.0	2.0
22:00	2 773	2 284	1 794	1 305	816	1.98	266.5	86.2	40.7	0.962	0.2	0.345	0.75	0.8	3.222	100	0.06	0.04	2.0	2.0
23:00	2 773	2 284	1 794	1 305	816	1.99	256.0	80.0	39.2	0.962	0.2	0.332	0.75	0.8	3.168	100	0.06	0.04	2.0	2.0
00:00	2 773	2 284	1 794	1 305	816	2.12	261.0	73.9	42.9	0.962	0.2	0.314	0.75	0.8	3.218	100	0.06	0.04	2.0	2.0
01:00	2 773	2 284	1 794	1 305	816	2.27	266.0	67.7	41.8	0.962	0.2	0.297	0.75	0.8	3.268	100	0.06	0.04	2.0	2.0
02:00	2 773	2 284	1 794	1 305	816	2.29	264.3	69.0	40.8	0.962	0.2	0.314	0.75	0.8	3.356	100	0.06	0.04	2.0	2.0
03:00	2 773	2 284	1 794	1 305	816	2.31	262.5	70.4	41.0	0.962	0.2	0.330	0.75	0.8	3.444	100	0.06	0.04	2.0	2.0
04:00	2 773	2 284	1 794	1 305	816	2.33	260.8	71.7	41.0	0.962	0.2	0.346	0.75	0.8	3.532	100	0.06	0.04	2.0	2.0
05:00	2 936	2 610	2 284	1 958	1 631	2.35	259.0	73.0	42.2	0.962	0.2	0.360	0.75	0.8	3.620	100	0.06	0.04	2.0	2.0
06:00	3 100	2 936	2 773	2 610	2 447	2.32	255.0	71.0	44.5	0.962	0.2	0.350	0.75	0.8	3.494	100	0.06	0.04	2.0	2.0

Note: BCOD = 0.52 COD, Total Bioreactor Volume = 1 000 m³, Sludge Blanket Volume = 100 m³

5.5.2.2 NO₃⁻-N removals in the anoxic zones and related energy savings

Figures 5.24(a) and (b) show diurnal NO_3^--N concentration profiles in the first and second anoxic compartments, respectively. A common feature was that the NO_3^--N concentration was low during the peak flow period and high during the base flow period. This tendency was more apparent with the higher peak factors. For example, with a peak factor of 1.2, the NO_3^--N concentration was 1.3 mg NO_3^--N l⁻¹ at 09:00 and 2.6 mg NO_3^--N l⁻¹ at midnight while with a peak factor of 2.0, the corresponding values were 0.5 and 7.9 mg NO_3^--N l⁻¹, respectively. The $BCOD/NO_3^-$-N ratio during the peak and low flow periods was a determining factor. With constant MLR and RAS ratios, the $BCOD/NO_3^-$-N ratio was high during the peak flow period when more carbon in the feed was available for reduction of NO_3^--N and resulted in enhanced denitrification. During the low flow period, NO_3^--N removal was reduced due to a lack of carbon as expressed by a low $BCOD/NO_3^-$-N ratio. The average NO_3^--N concentration of the second anoxic compartment (Figure 5.24(b)) was 1.8 mg NO_3^--N l⁻¹ lower than that of the first anoxic compartment (Figure 5.24(a)), and almost similar to 1.9 mg NO_3^--N l⁻¹, the value of the laboratory findings (Section 4.3.4.1). This clearly indicated enhanced denitrification in the second anoxic compartment.

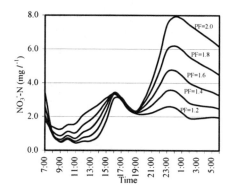 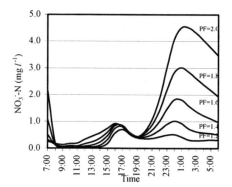

Figure 5.24(a). Diurnal NO_3^--N concentration profiles of the first anoxic compartment with different peak factors.

Figure 5.24(b). Diurnal NO_3^--N concentration profiles of the second anoxic compartment with different peak factors.

Figures 5.25(a) and (b) show the diurnal NO_3^--N removals in the first and second anoxic zones, which were calculated using Equation 3.1 for different peak factors. Table 5.17 shows the average NO_3^--N removals that corresponded with the different peak factors in the first and second anoxic zones. The overall average NO_3^--N removal in the first anoxic zone was 19.5 mg NO_3^--N l⁻¹ while that in the second anoxic zone was 5.6 mg NO_3^--N l⁻¹.

The results showed that due to the expansion of the anoxic volume, denitrification was enhanced by 29% based on the NO_3^--N removal in the first anoxic zone, which was consistent with the experimental results of Section 4.3.4.1.

Table 5.17. Average NO_3^--N removed, in terns of the mass removal rate and influent-based concentration, in anoxic zones 1 and 2 with different peak factors.

Peak factor	NO_3^--N removal			
	Mass removal rate (kg d^{-1})		Influent-based concentration (mg N l^{-1})	
	Anoxic zone 1	Anoxic zone 2	Anoxic zone 1	Anoxic zone 2
1.2	62.3	17.3	20.0	5.3
1.4	65.0	17.9	19.9	5.5
1.6	64.0	18.6	19.6	5.7
1.8	63.0	18.6	19.3	5.7
2.0	61.3	18.3	18.8	5.6
Average	63.1	18.1	19.5	5.6

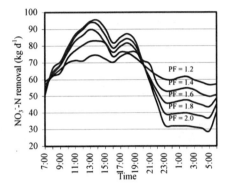 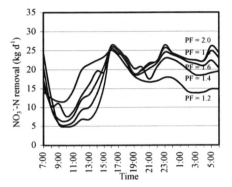

Figure 5.25(a). Diurnal mass removal rate of NO_3^--N in the first anoxic compartment with different peak factors.

Figure 5.25(b). Diurnal mass removal rate of NO_3^--N in the second anoxic compartment with different peak factors.

Table 5.18 shows the oxygen demands for COD removal and nitrification, and the oxygen demand reductions due to denitrification. The calculations were made from the sewage characteristics used in this simulation, the calculated average TKN oxidized and the NO_3^--N removed in the anoxic zones. The average process yield coefficient of sludge production was 0.32 g COD (g COD$_{REM}$)$^{-1}$, which was based on the COD removal and sludge production of the optimized laboratory-scale MLE system (Section 4.3.7.1). The oxygen consumption per mg NH_4^+-N oxidized and the oxygen equivalent per mg NO_3^--N denitrified, were 4.57 and 2.86 mg O_2, respectively (Rittmann and McCarty, 2001). Table 5.19 shows the oxygen demands for COD removal and nitrification, the total oxygen demands for both and the total oxygen demand reductions for three different configurations:

i   COD and ammonia removals in the conventional activated sludge processes;

ii   COD and nitrogen removals in the existing MLE process with an anoxic volume of 25%; and

iii COD and nitrogen removals in the recommended MLE process with an anoxic volume of 50%.

The oxygen demand reduction due to denitrification was calculated based on the conventional process.

Table 5.18. Oxygen demands due to COD removal and nitrification, and oxygen demand reductions due to denitrification.

Item	Unit	COD removal	Nitrogen removal		
				Denitrification	
			Nitrification	25% anoxic volume	50% anoxic volume
COD	mg COD l^{-1}	270	-	-	-
TKN	mg N l^{-1}	-	42.5	-	-
NO_3^--N	mg N l^{-1}	-	-	19.5	25.1
Oxygen demand	mg O_2 l^{-1}	184 (270 x (1 - 0.32[*]))	194 (42.5 x 4.57[**])	-	-
Oxygen demand reduction	mg O_2 l^{-1}	-	-	56 (19.5 x 2.86[***])	72 (25.1 x 2.86[***])

[*] Process sludge yield (Y) calculated from the data of the optimized laboratory-scale MLE system of Phase III of Kranji WRP

[**] Oxygen consumption per mg NH_4^+-N oxidized

[***] Oxygen equivalent per mg NO_3^--N denitrified

Table 5.19. Oxygen demands and oxygen demand reduction of the activated sludge processes for the three different configurations.

Option	Oxygen demand (mg O_2 l^{-1})			Total oxygen demand reduction (%)
	COD removal	Nitrification	Total	Total
COD and ammonia removals in the conventional activated sludge processes	184	194	378	0.0
COD and nitrogen removal in the MLE process with 25% anoxic volume	128	194	322	14.8
COD and nitrogen removal in the MLE process with 50% anoxic volume	112	194	306	19.0

The oxygen demand reduction in the first anoxic compartment was 14.8% (Table 5.19), which was comparable with the practical and laboratory values. The oxygen demand reduction of Phase IV of Bedok WRP is 9% (Figure 3.20) and 18% for the laboratory simulation, in which acetate was added to the feed (Figure 4.16). The total oxygen demand reduction for the first and second compartments was 19% (Table 5.19), comparable with 16% for the laboratory experiment with 50% anoxic volumetric ratio (Section 4.3.7.1). These data illustrated the benefits of oxygen demand reduction by expanding the existing anoxic volume from 25% to 50%. In reality, an energy saving of up to 33% may be achievable by omitting the aeration in the second compartment of the current system provided that the present aeration supply to the following aerobic compartments was sufficient to satisfy the oxygen demands of NH_4^+-N and COD transferred from the preceding expanded anoxic compartment.

5.5.2.3 Diurnal heterotrophic and autotrophic nitrifier concentrations

Figures 5.26(a) and (b) show the diurnal concentration profiles of heterotrophic biomass in the first and second anoxic compartments, respectively, while Figures 5.27(a) and (b) show the diurnal concentration profiles of autotrophic nitrifier in the first and second aerobic compartments, respectively. A common feature was that the biomass concentrations reached their lowest values during peak flow and their highest values during low (base) flow. The differences were as high as ±18% of the average biomass concentration in the compartment. An inverse tendency was found for the biomass concentration profile in the RAS (Figure 5.27(c)) where the concentration of heterotrophic biomass increased during the peak flow period. It appeared that the fluctuations in the compartments were largely due to the influent hydraulic flow (and MLR), which overrode the effect of high biomass concentration return from the RAS. However, calculation indicated that the total amount of heterotrophic biomass in the system did not increase during the peak flow as time interval of 2–3 h under the increased mass load was too short for heterotrophic growth, and even the instantaneous SRT during the peak flow could be reduced due to MLSS reduction and sludge overflow (Cao *et al.*, 2004c). During the base flow at night, the biomass concentrations in the compartments were more 'concentrated' than 'diluted' with the influent and MLR. The differences in the biomass concentrations between the peak flow and the base flow were more pronounced as the peak factor increased. This was similar to what was observed during wet weather where reduced MLSS concentration in the activated sludge tanks flowed into the clarifier, although biomass fluctuations in an actual system might be less pronounced than in the simulation due to hydraulic barriers in the system.

A decrease between 5 and 15 mg COD l^{-1} in the heterotrophic biomass concentration was found between the first and second anoxic compartments, which was coincident with an autotrophic biomass concentration decrease of between 2 and 5 mg COD l^{-1}. These differences were possibly due to growth exceeding decay in the first reactor. The autotrophic nitrifier and heterotrophs comprised 13% and 87% of the total active biomass in the system, respectively.

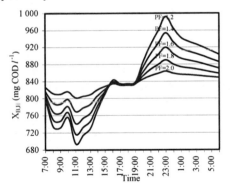

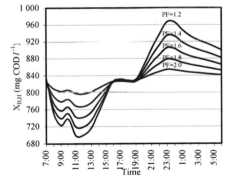

Figure 5.26(a). Diurnal concentration profiles of heterotrophic biomass in the first anoxic compartment with different peak factors.

Figure 5.26(b). Diurnal concentration profiles of heterotrophic biomass in the second anoxic compartment with different peak factors.

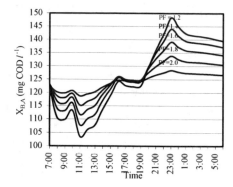

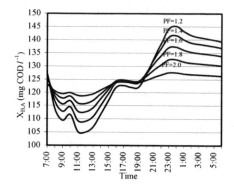

Figure 5.27(a). Diurnal concentration profiles of autotrophic nitrifier in the first aerobic compartment with different peak factors.

Figure 5.27(b). Diurnal concentration profiles of autotrophic nitrifier in the second aerobic compartment with different peak factors.

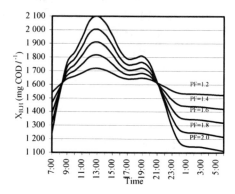

Figure 5.27(c). Diurnal concentration profiles of heterotrophic biomass in the RAS with different peak factors.

The fluctuation of biomass concentration could have an impact on the effluent quality, particularly the NH_4^+-N concentration, especially when the peak flow coincided with the peak NH_4^+-N mass load. The reduction of autotrophic nitrifier concentration during the peak flow lowered the nitrification capacity and, thus, increased the effluent NH_4^+-N concentration as shown in Figures 3.7 and 4.6, and through simulation in Figures 5.28(a) and (b). In contrast, the nitrification capacity of the relatively higher biomass concentration was sufficient for the ammonia load during base flow at night. Therefore, adoption of a high RAS ratio during the peak flow period and a low RAS ratio during the base flow period could minimize the effluent NH_4^+-N fluctuations.

5.5.2.4 Diurnal NH_4^+-N and NO_3^--N concentration profiles of the aerobic compartments

As shown in Figures 5.28(a) and (b), the maximum NH_4^+-N concentration of the first aerobic compartment increased from 5.3 to 13.3 mg NH_4^+-N l^{-1} as the peak factor increased from 1.2 and 2.0 attributing to the high NH_4^+-N mass load and low nitrifier concentration during the peak flow period. This tendency was attenuated in the second aerobic compartment where the maximum NH_4^+-N concentrations reduced to 0.3 and 5.0 mg NH_4^+-N l^{-1} for peak factors of 1.2 and 2.0, respectively. The reduced NH_4^+-N mass load was supposedly the reason.

As shown in Figure 5.28(b), the average NH_4^+-N concentration for peak factor of 2.0 was < 5.0 mg NH_4^+-N l^{-1}, the feed water requirement for NEWater production. Therefore, a five-day aerobic SRT should be sufficient to produce stable effluent quality even under strong hydraulic flow fluctuations.

The NO_3^--N concentration profiles in the first aerobic compartment, as shown in Figure 5.29(a), fell between 10:00 and 16:00, corresponding to the NH_4^+-N concentration increases in the same compartment during the same period. However, this effect was much less apparent in the second aerobic compartment, as shown in Figure 5.29(b), corresponding to less NH_4^+-N concentration increases during the peak flow periods.

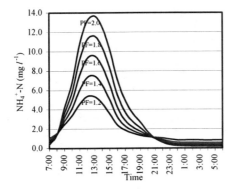

Figure 5.28(a). NH_4^+-N concentration profiles of the first aerobic compartment with different peak factors.

Figure 5.28(b). NH_4^+-N concentration profiles of the second aerobic compartment with different peak factors.

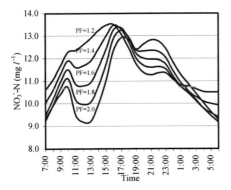
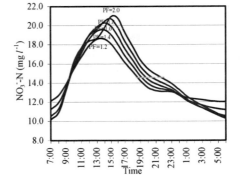

Figure 5.29(a). Diurnal NO_3^--N concentration profiles of the first aerobic compartment with different peak factors.

Figure 5.29(b). Diurnal NO_3^--N concentration profiles of the second aerobic compartment with different peak factors.

5.5.2.5 Diurnal oxygen consumption in the aerobic compartments

Figures 5.30(a) and (b) show the diurnal oxygen consumption in the first and second aerobic compartments where a DO concentration of 2 mg O_2 l^{-1} was maintained in each. The consumption was high during the day, which was coincident with the peak NH_4^+-N mass load. The consumption in the first aerobic compartment was higher and prevailed longer than in the second aerobic compartment.

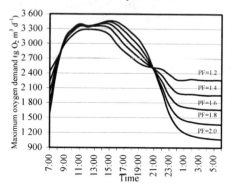
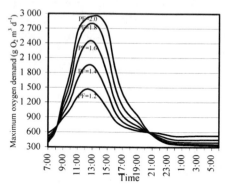

Figure 5.30(a). Diurnal oxygen consumption profiles of the first aerobic compartment with different peak factors.

Figure 5.30(b). Diurnal oxygen consumption profiles of the second aerobic compartment with different peak factors.

Considering the fact that only a limited portion of oxygen supply in the first aerobic compartment was used up for COD removal as the bulk of COD was removed in the two preceding anoxic reactors through denitrification, the oxygen consumption in the two aerobic compartments, therefore, mirrored the NH_4^+-N mass loads at the respective inlets. Thus, the oxygen consumption was governed, to a large extent, by the magnitude and retention time of the high NH_4^+-N mass loads to the compartments. The difference in the magnitude of oxygen consumption during the peak periods in each of the two aerobic compartments was not significant, but the retention time in the first aerobic compartment was much longer than the second. The difference between the oxygen consumption during the day and that at night within the same compartment was significant, especially for the higher peak factors. The oxygen consumption in the first aerobic compartment during the day was 1.5 and 3 times greater than the consumption at night for peak factors of 1.2 and 2.0, respectively. For the second aerobic compartment, the equivalent consumption was 3 and 10 times greater than the consumption at night. Thus, more air should be supplied during the day while the aeration could be reduced at night to save energy significantly.

Given the performance of the full-scale systems, the optimal laboratory simulation and mathematical simulation performed, the recommended design guidelines for the BNR activated sludge process in warm climates with regard to the SRT, HRT and volumetric ratios of the anoxic and aerobic compartments are proposed. In fact, at the time of writing, the findings of this study have been put into practice in full-scale application. The activated sludge trains of Battery B of Seletar WRP are now operated based on the process configuration and operation proposed in this study. The designs of several new plants in Singapore are also similar to the recommendations in this study. For example, Changi WRP, which is one of the largest in Asia and is under commissioning at the time of writing, adopts an anoxic volumetric ratio of 50% and an aerobic SRT of 4 days; and in a full-scale membrane bioreactor (MBR) plant, which is under construction, an anoxic volumetric ratio of 40% is adopted.

5.6 SUMMARY

COD-based influent model

A COD-based influent model developed based on the results of sewage characterization under Singapore conditions allows the use of COD, SCOD, TSS, VSS, TKN and NH_4^+-N analyses in the regular monitoring/sampling programme, together with the suggested and calculated conversion coefficients, as data for the influent model in modeling with ASM No. 1. The results of parameter identification indicated that the methods suggested for the determination of the conversion coefficients of the COD-based influent model were workable. This largely simplifies the characterization work and eases the application of modeling and simulation under Singapore conditions.

Parameter identification

Calibration. The kinetic parameters and stoichiometric coefficients of heterotrophs and autotrophic nitrifiers were first calibrated with the measured data obtained from the laboratory experiment of the MLE process of Phase IV of Bedok WRP. Only three parameters, $\hat{\mu}_A$, anoxic Y_{HD}, and b_H, were regulated during the calibration, and the other parameters were taken either from the default values or the values after temperature correction. These parameters were applied successfully to describe a laboratory-scale MLE process under the dynamic feed conditions of Phase IV of Bedok WRP.

Verification. The calibrated kinetic parameters and stoichiometric coefficients were verified separately with five measured data sets from different activated sludge processes of both laboratory- and full-scale processes of the three different WRPs under dynamic feed conditions.

In these verifications, the general fits between the simulated and measured NH_4^+-N and NO_3^--N concentration profiles were satisfactory. No further calibration and verification of parameter values were required. The kinetic constants and coefficients of ASM No. 1 were applicable for modeling BNR in MLE and other activated sludge processes such as simultaneous nitrification and denitrification (SND), Bardenpho and step-feed activated sludge processes in several WRPs under Singapore conditions.

Design of the BNR activated sludge process in warm climates

An aerobic SRT of about 4 days for nitrification can be adopted in the design of BNR activated sludge process in warm climates corresponding to a minimum aerobic SRT of about 2 days. The process aerobic SRT is about half of that of the existing activated sludge process. An anoxic SRT of between 3 and 5 days is proposed for denitrification. It is longer than that of the conventional design because the concentration ranges of both the readily biodegradable COD and alkalinity of the sewage in Singapore are near the low boundary of the respective ranges of reported values. The final selection of anoxic SRT depends on the sewage characterization, the higher the total COD and readily biodegradable COD, the shorter the anoxic SRT while the lower the total COD and readily biodegradable COD, the longer the anoxic SRT.

Design for upgrading the existing BNR activated sludge process in warm climates

A MLE process with an anoxic and an aerobic volumetric ratio of 50% each is the recommended configuration proposed for the existing activated sludge process to be upgraded to. With respect to the existing activated sludge process, the current anoxic volume is to be expanded from 20-25% to 50% while the current aerobic volume is to be reduced from 75-80% to 50%. The total SRT of 10 d and the HRT of 6 h, which are similar to those of the existing process, can be retained. Both MLR and RAS ratios are 100% each but smaller ratios can be adopted during the low (base) flow period.

The simulations identified the advantages of expanding the anoxic volume, reducing the aerobic volume and adopting the following operational strategies:

i   denitrification was enhanced by 29% with the increased anoxic volume;

ii   the daily average NH_4^+-N concentration was $<<$ 5 mg NH_4^+-N l^{-1} for an aerobic SRT of 5 d, even with a peak flow factor of 2.0;

iii the biomass concentrations in the reactors reduced during the peak flow period and increased during the base flow period. This biomass reduction could cause the effluent NH_4^+-N concentration to increase during the peak flow period. Adoption of a high RAS ratio during the peak flow period and a low RAS ratio during the base flow period could minimize the effluent NH_4^+-N concentration fluctuations; and

iv   the air supply to each compartment could be regulated according to the oxygen demand during the day and at night to save aeration energy further.

Implications for the process development of the activated sludge process

The difference in scale between the laboratory- and full-scale systems is in the order of magnitude of 1,000,000 times i.e., 5.6 litre versus 8 724 m^3. The magnitude difference is the largest ever reported in the scale-down of laboratory experiments and in scale-up modeling to study full-scale activated sludge processes. The success in parameter identification illustrated that:

i   the methodologies and approaches of the study (Figure 1.2) were successful;

ii   the scale-down principles developed in the design of a laboratory experiment to investigate a full-scale activated sludge process were workable; and

iii the approach to calibrate a few model parameters obtained from a laboratory experiment to predict the performance of a full-scale system was feasible.

Given the technology advancements to date, there are no principal difficulties in establishing a laboratory-scale system, at a reasonable cost, to simulate a full-scale activated sludge process under dynamic feed conditions, even with the incorporation of daily and seasonal temperature changes, non-ideal mixing and dead space in full-scale process into the design. The outcomes of this study are encouraging and demonstrate the feasibility to adopt a laboratory experiment to study an existing full-scale system and design a new process. "From Flask to Field" appears to be a real possibility rather than purely a dream in activated sludge wastewater treatment research and development, which could save a lot of time and resources.

REFERENCES

Cao Y.S. and Alaerts G.J. (1995) Influence of Reactor Type and Shear Stress on Aerobic Biofilm Population, Morphology and Kinetics. Wat. Res. **29**(1), 107-118.

Cao Y.S., Ang C. M. and Zhao W. (2003) Performance Analysis of Laboratory and Full-Scale Activated Sludge Process in Kranji and Bedok Water Reclamation Plants. Technical Report, Ref no. SUI/2001/030/TRTP1.

Cao Y.S., Ang C.M. and Raajeevan K.S. (2004a) Laboratory-Scale Studies of the Activated Sludge Process of Seletar Water Reclamation Plant. Technical Report, Ref no. SUI/2001/030/TRTP4.

Cao Y.S., Raajeevan K.S. and Ang C.M. (2004b) Optimization of Biological Nitrogen Removal in Municipal Wastewater Treatment by the Activated Sludge Process in Singapore. Technical Report, Ref no. SUI/2001/030/TRTP2.

Cao Y.S., Raajeevan K.S. and Ang C.M. (2004c) Modeling Biological Nitrogen Removal in Activated Sludge Process in Singapore. Technical Report, Ref no. SUI/2001/030/TRTP3.

Cao Y.S., Raajeevan K.S. and Ang C.M. (2004d) Simulation of the Step-Feed Activated Sludge Process of Phase I of Changi WRP, Technical Report, Ref no. SUI/2001/030/TRTP5.

Carucci A., Rolle E. and Smurra P. (1999) Management Optimization of a Large Wastewater Treatment Plant. Proc. Application of Models in Water Management, Amsterdam, 249-256.

Chen M.H. (1987) Process Modeling. Eastern University of Science and Technology, Shanghai, China.

Concha L. and Henze M. (1992) Advanced Design and Operation of Municipal Wastewater Treatment Plant. Technologies for Environmental Protection Report 1, EUR 16869 EN.

Concha L. and Henze M. (1996) Advanced Design and Operation of Municipal Wastewater Treatment Plant. Technologies for Environmental Protection Report 10, EUR 16869 EN.

Daigger G.T. and Nolasco D. (1995) Evaluation and Design of Full-Scale Wastewater Treatment Plants Using Biological Process Models. Wat. Sci. Tech. 31(2), 245-255.

De La Sota A., Larrea L., Novak L., Grau P. and Henze M. (1994) Performance and Model Calibration of R-D-N Process in Pilot Plant. Wat. Sci. Tech. 30(6), 355-364.

Dold P.L., Ekama G.A. and Marais G.v.R. (1980) A General Model for the Activated Sludge Process. Prog. Wat. Tech., 12(6), 47-77.

EPA (1993) Manual Nitrogen Control, EPA/625/R-93/010, Washington D.C.

Grady C.P.L. (1992) Environmental Biotechnology: From Flask To Field, in Advanced Course on Environmental Biotechnology. Delft University of Technology, May 6-14, 1993, Delft, The Netherlands.

Gragy C.P.L., Daigger G.T. and Lim H.C. (1999) Biological Wastewater Treatment, 2nd ed., Marcel Dekker. New York.

Gujer W., Henze M., Mino T. and van Looschrecht M.C.M. (1999) Activated Sludge Model No. 3. Wat. Sci. Tech. 39(1), 183-193.

Harromoës P., Haarbo A., Winther-Nielsen M. and Thirsing C. (1998) Six Years of Pilot Studies for Design of Treatment Plants for Nutrient Removal. Wat. Sci. Tech. 38(1), 219-226.

Henze M., Grady C. P., Gujer W., Marais G. v. R. and Matsuo T. (1987). Activated Sludge Model No. 1. IAWPRC Sci. and Tech Report No. 1, IAWPRC, London.

Henze M. (1992) Characterization of Wastewater for Modeling of Activated Sludge Processes. Wat. Sci Tech. 25(6), 1-15.

Henze M., Grady C.P.L., Gujer W., Marais G.v.R. and Matsuo T. (1987) Activated Sludge Model No. 1. Scientific and Technical Report No. 1, IAWPRC, London.

Henze M., Gujer W., Mino T., Matsuo T., Wenzel M.C. and Marais G. v. R. (1995), Wastewater and Biomass Characterization for the Activated Sludge Model No. 2: Biological Phosphorus Removal. Wat. Sci. Tech. **31**(2), 13-23.

Henze M., Gujer W., Mino T., Matsuo T., Wentzel M.C., Marais G.v.R. and van Looschrecht M.C.M. (1999) Activated Sludge Model No. 2d, ASM2D. Wat. Sci. Tech. **39**(1), 165-182.

Hulsbeek J.J. W., Kruit J., Roeleveld P.J. and van Loosdrecht M.C.M. (2002) A Practical Protocol for Dynamic Modeling of Activated Sludge Systems. Wat Sci. Tech. **45**(6), 127-136.

Hydromantis Inc. (1999) Technical Reference of GPS-X, Version 3.0. Canada.

Kalker T.J.J., van Guor G.P., Roeleveld P.J., Ruland M.F. and Babugka R. (1999) Fuzzy Control of Aeration in an Activated Sludge Wastewater Treatment Plant: Design, Simulation and Evaluation. Wat. Sci. Tech. **39**(4), 71-78.

Kappeler J. and Gujer W. (1992) Estimation of Kinetic Parameters of Heterotrophic Biomass Under Aerobic Conditions and Characterization of Wastewater for Activated Sludge Modeling. Wat. Sci. Tech. 25(6), 125-139.

Kossen N.W.F. and Oosterhuis N.M.G. (1985) Modeling and Scaling-Up of Bioreactor. In Biotechnology, (eds. Rehm H. J. and Reed D.), Vol.2, 572-605, VCH Verlaggesllschaft, Weinheim.

Kruit J. and van Loosdrecht M.C.M. (2002) A Practical Protocol for Dynamic Modelling of Activated Sludge Systems. Wat. Sci. Tech. 45(6), 127-136.

Ladiges G., Gunner C.H. and Otterpohl R. (1999) Optimization of the Hamburg Wastewater Treatment Plant by Dynamic Simulation. Wat. Sci. Tech., 39(4), 37-45.

Langeveld J. G. (2004) Interactions Within Wastewater Systems, PhD Dissertation, Delft University of Technology, The Netherlands.

Lessouef A., Payraudeau M., Regalla F. and Kleiber B. (1992) Optimizing Nitrogen Removal Reactor Configurations by On-Site Calibration of the IAWPRC Activated Sludge Model. Wat. Sci. Tech.25(6), 105-123.

Makinia J., Swinarski M. and Dobiegala E. (2002) Experiences with Computer Simulation at Two Large Wastewater Treatment Plants in Northern Poland. Wat. Sci. Tech. 45(6), 209-218.

Marais G.v.R. and Ekama G.A. (1976) The Activated Sludge Process: Part 1 – Steady State Behaviour. Wat. SA 2, 164-200.

McKinney R.E. and Ooten R.J. (1969) Concepts of Complete Mixing Activated Sludge. Trans. 19th Eng. Conf., Univ. of Kansas, 32.

Meijer S.C.F. (2004) Theoretical and Practical Aspects of Modeling Activated Sludge Processes, PhD Thesis, Delft University of Technology, The Netherlands.

Melcer H. (1999) Full-Scale Experience with Biological Process Models-Calibration Issues. Wat. Sci. Tech. 39(1), 245-252.

Metcalf & Eddy Inc. (2003), Wastewater Engineering: Treatment and Reuse. 4th ed., McGraw-Hill, Washington, USA.

Orhon D., Sozen S. and Artan A. (1996) The Effect of Heterotrophic Yield on the Assessment of the Correction Factor for Anoxic Growth. Wat. Sci. Tech. 34(5/6), 67-74.

Raajeevan K.S. (2003) Biological Nitrogen Removal in Wastewater Treatment Plants in Singapore. M.Eng. Thesis. National University of Singapore

Rittmann B.E. and McCarty P.L. (2001) Environmental Biotechnology: Principles and Applications. McGraw-Hill, Singapore.

Roeleveld P.J. and van Loosdrecht M.C.M. (2002) Experiences with Guidelines for Wastewater Characterization in the Netherlands. Wat. Sci. Tech. 45(6), 77-88.

Salem S., Berends D., van Loosdrecht M.C.M. and Heijnen J.J. (2002). Model Based Evaluation of A New Upgrading Concept for N-removal. Wat. Sci. Tech. 45(6), 169-176.

Siegrist H. and Tschui M. (1992) Interpretation of Experimental Data with Regard to the Activated Sludge Model No. 1 and Calibration of the Model for Municipal Wastewater Treatment Plants. Wat. Sci. Tech. 25(6), 167-183.

Spanjers H. (1993) Respirometry in Activated Sludge. PhD Thesis, Wageningen University, The Netherlands.

Takàcs I. (2006) Personal Communication.

van Haandel A.C., Ekama G.A. and Marais G.v.R. (1981) The Activated Sludge Process Part 3 – Single Sludge Process. Wat. Res. 15, 1135-1152.

Van Veldhuizen H.M., van Looschrect M.C.M. and Brandse F.A. (1999) Model Based Evaluation of Plant Improvement Strategies for Biological Nutrient Removal. Wat. Sci. Tech. 39(4), 45-55.

Vanrolleghem P.A., Spanjers H., Petersen B., Ginestet P. and Takacs I. (1999) Estimating (Combinations of) Activated Sludge Model No. 1 Parameters and Components by Respirometry. Wat. Sci. Tech. 39(1), 195-214.

WRc plc (1998) STOAT: Dynamic Modeling of Wastewater Treatment Works. Technical Document. Wiltshire, UK.

WERF (2003) Methods for Wastewater Characterization in Activated sludge Modeling. WERF Report 99-WWF-3, Co-published by IWA Publishing and the Water Environment Federation.

6

Summary

6.1 BACKGROUND AND APPROACHES OF THE STUDY

To cope with the feed water quality requirements of the NEWater plants, upgrading of the activated sludge processes by incorporating biological nitrogen removal (BNR) into the existing activated sludge processes was undertaken at three water reclamation plants (Bedok, Kranji and Seletar WRPs) in Singapore, a country with a warm climate. The principal objective of the identified study was 'to ensure the secondary effluent meets the required quality specifications through the study of the BNR activated sludge processes in the three WRPs, and to develop the design guidelines for the BNR activated sludge process in warm climates'. The specific objectives of the study were further refined as follows:

i   To evaluate the design of the upgraded activated sludge processes in the three WRPs and to provide information that is helpful in the operation of the BNR activated sludge process;

ii   To explore the feasibility of using a laboratory-scale system for the simulation and study of the performance of the full-scale activated sludge process. Efforts were made to explore the possibility of from flask to field i.e., using a 5.6 litre laboratory-scale activated sludge system for the simulation and prediction of the performance of the full-scale activated sludge process with a volume that ranged between 6 000 and 8 700 m^3; and

iii To make recommendations on the optimal upgrading of the BNR activated sludge process in warm climates and develop a capacity that enables simple, reliable and modeling-aided design of BNR activated sludge processes in warm climates.

© IWA Publishing 2008. *Biological Nitrogen Removal Activated Sludge Process In Warm Climates: Full-scale process investigation, scaled-down laboratory experimentation and mathematical modeling* by Cao Ye Shi, Wah Yuen Long, Ang Chee Meng and Kandiah S. Raajeevan. Published by IWA Publishing, London, UK. ISBN: 1843391872.

Holistic and systematic approaches have been applied in this study. In essence, three blocks constituted the main work:

i   full-scale investigation whereby detailed investigation on the design, performance and efficiency of the whole activated sludge process as well as the individual compartments, under site conditions, was carried out. Characterization of the settled sewage and sludge and 24-h sampling programme helped in understanding the process performance and preparation of the feed adopted in the laboratory experiment and development of the influent model in modeling;

ii   laboratory experimentation, whereby the scale-down principles in the activated sludge process were developed and the parameters of 'regime analysis' were identified. The laboratory-scale system designed accordingly was established and employed for simulation of the performance of the site process; and

iii   mathematical modeling, which included development of a simple influent input model that allows the use of regular monitoring data in modeling, parameter identification by using the measured data obtained from the laboratory experimentation and full-scale investigation, simulation of the performance under various site conditions, and development of the guidelines for upgrading and design of the BNR activated sludge process in warm climates.

The major findings are summarized in the following sections.

6.2 CHARACTERIZATION OF THE SETTLED SEWAGE AND ACTIVATED SLUDGE

Diurnal hydraulic flow. For the three WRPs investigated, at least one peak flow with a factor of about 1.5 occurred in the day and a base flow with a factor ranging between 0.2 and 0.6 occurred between midnight and early morning. For some WRPs, another peak flow was observed in the evening. The COD and NH_4^+-N mass loading rates coincided with the respective first flow peak and base flow.

Conventional parameters. The SCOD values of the settled sewages of the three WRPs were 100, 151 and 117 mg l^{-1} respectively, which were within the 'diluted' ranges while the particulate COD concentrations of the settled sewages were high. The alkalinity values of the settled sewages of the three WRPs were 217, 179 and 148 mg l^{-1} (as $CaCO_3$) respectively, which were closer to the lower boundary of reported range. The ALK/NH_4^+-N ratio of 5.9 mg $CaCO_3$ (mg NH_4^+-N)$^{-1}$ was < 7.08 mg $CaCO_3$ (mg NH_4^+-N)$^{-1}$, and SCOD/TKN ratio was close to or < 2.86 mg COD (mg NO_3^--N)$^{-1}$.

Diluted SCOD, low alkalinity and low ratios of ALK/NH_4^+-N and SCOD/TKN indicated that:

i   the soluble biodegradable COD might not be sufficient for efficient nutrient removal;

ii   either alkalinity addition or denitrification would be necessary when nitrification is facilitated; and

iii   a large anoxic reactor volume might be required for a high denitrification efficiency. The COD/BOD_5 ratios of the settled sewages of the three WRPs indicated a high portion ($\geq 20\%$) of inert solids in the settled sewages.

COD fractions. The COD fingerprints varied with the hydraulic flow. The readily biodegradable COD (S_S) of the settled sewage of Bedok WRP during the peak, normal and low flow periods were 66, 51 and 6 mg l^{-1}, respectively; and the corresponding values were 36, 50 mg l^{-1} and non-detectable for that of Phase III of Kranji WRP. The S_S was almost depleted during the low flow due to the longer sewage retention time. The rapidly hydrolysable biodegradable COD (X_{S1}) of the settled sewage of Phase IV of Bedok WRP during the three flow periods were 60, 27 and 28 mg l^{-1}, respectively; and the corresponding values were 81, 49 and 90 mg l^{-1} for Phase III of Kranji WRP. These data indicated that the diurnal COD fractions, rather than the fixed COD fractions, should be used in modeling and simulations.

Compared to the literature data, the S_S of the settled sewage, which varied between non-detectable and 66 mg l^{-1}, was regarded as "very diluted" and lower than reported values. The elevated temperature of the warm climate could be the main reason.

Denitrification rates corresponding to COD fractions. The average specific rate corresponding to S_S (r_{DSS}) was 7.20 mg N (g VSS)$^{-1}$ h^{-1}. The specific rate corresponding to X_{S1} (r_{DX1}) was 1.80 mg N (g VSS)$^{-1}$ h^{-1}. The specific rates corresponding to X_{S2} (r_{DX2}) and X_B (r_{Dend}) were 0.63 and 0.45 mg N (g VSS)$^{-1}$ h^{-1}, respectively.

Spatial distribution of denitrification and nitrification activities. The specific nitrification and denitrification activities of the activated sludges of the aerobic and anoxic compartments of Phase IV of Bedok WRP had no significant changes. This demonstrated that the active nitrifier and denitrifier compositions in the sludges of the anoxic and aerobic compartments of a single sludge activated sludge process were comparable, and verified the rationale to adopt the same set of kinetics of nitrification and denitrification in the aerobic and anoxic compartments of a single sludge activated sludge process in modeling.

6.3 PERFORMANCE OF THE FULL-SCALE ACTIVATED SLUDGE PROCESS

Key parameters. The key design parameters including the configuration, sizes, SRT, HRT, and ratios of MLR and RAS etc., of the MLE activated sludge process were collected through (i) the design manual and (ii) study of the plants' records. The SRT was verified through the calculation based on the statistical data of COD mass load, MLSS and wasting rate.

COD removal. The average SCOD values in the settled sewage, the anoxic and the three aerobic compartments of the activated sludge process of Phase IV of Bedok WRP were 109, 24, 14, 18 and 11 mg l^{-1}, respectively. Most of the SCOD was removed in the anoxic compartment due to denitrification, while SCOD removals in the second and third aerobic compartments were insignificant. The average SCOD removal efficiency was 90.0%.

Nitrification. The average diurnal NH_4^+-N concentrations in the settled sewage and the four compartments were 25.6, 10.3, 2.0, 1.4 and 1.6 mg NH_4^+-N l^{-1}, respectively, indicating that most of the NH_4^+-N in the feed was removed in the first aerobic compartment, and the contributions of the second and third compartments in NH_4^+-N oxidation were marginal. The average NH_4^+-N removal efficiency was 93.8 %. The NH_4^+-N concentration profiles in the three aerobic compartments were « 5 mg NH_4^+-N l^{-1}, the feed water quality requirement of the NEWater plants. The specific nitrification activity correlated well with the ambient NH_4^+-N mass load.

Denitrification. The average NO_3^--N concentrations in the four compartments were 1.2, 6.9, 8.2 and 8.9 mg NO_3^--N l^{-1}, respectively. The average NH_4^+-N-based denitrification efficiency was 62.1%. The anoxic compartment contributed 34.0%, the final clarifier sludge blanket contributed 23.0% and the aerobic compartments contributed 5.1%.

COD balance. The percentage distribution of the influent COD in descending order in the activated sludge process was: 47%, oxygen-COD consumption in the aerobic compartments; 28%, COD assimilation into the biomass; 9%, COD removal in the anoxic compartment; 7%, COD removal in the final clarifier; 7%, particulate COD in the final effluent; and 2%, soluble COD in the final effluent. The net sludge yield was 30.2%. The total COD catabolism coefficient was 69.8%. A 26% reduction on carbonaceous oxygen requirement was achieved due to denitrification in the anoxic zone and final clarifier. Taking into account the oxygen requirement for autotrophic ammonia–nitrogen oxidation, a 12% total oxygen demand reduction was achieved.

Nitrogen balance. The percentage distribution of influent nitrogen in descending order in the activated sludge process was: 26%, NO_3^--N removal in the anoxic compartment; 24%, NO_3^--N removal in the final effluent; 20%, NO_3^--N removal in the final clarifier; 18%, nitrogen assimilation into sludge cells; 5%, NH_4^+-N in the final effluent; 4%, suspended TKN in the final effluent; and 3%, soluble organic nitrogen in the final effluent.

This detailed information and investigation of the diurnal feed, carbonaceous and nitrogenous matter conversion in individual parts as well as of the whole full-scale activated sludge process exhibited a clear, dynamic picture of the system performance, and helped in giving an insightful understanding of the process. These data were used in the design and study of the laboratory-scale process and the parameter verification of modeling.

6.4 SCALED-DOWN LABORATORY INVESTIGATION

Scale-down principle for activated sludge process. Three categories of the 'ruling regimes' for the scale–down of the activated sludge process are defined in this study:

i   feed conditions, including diurnal hydraulic flow, sewage compositions based on both conventional parameters and COD fractions;

ii   bioreactor system, including hydraulic flow pattern either ideal or non–ideal, configuration and sizes etc.; and

iii   biochemical environment, including SRT, volumetric ratios of anaerobic/anoxic and aerobic zones in the activated sludge process; MLVSS concentrations, wasting rates, recycling ratios, HRT, temperature, DO concentrations in the systems. The laboratory-scale process should share similar ruling regimes as those of the full-scale process investigated.

Simulation of the activated sludge process of Phase IV of Bedok WRP. The results of the laboratory-scale studies indicated that the bulk of the SCOD and nitrate-nitrogen reduction could have occurred in the anoxic reactor. Ammonia-nitrogen oxidation was mainly in the first aerobic reactor, and the contribution of the second aerobic reactor was marginal. The SCOD, NH_4^+-N, and NO_3^--N concentration profiles were quite similar to those of the full-scale activated sludge process of Bedok WRP. The ranges of the specific rates of nitrification and denitrification were similar to those of the full-scale activated sludge process of Bedok WRP too.

The SCOD, NH_4^+-N, NO_3^--N and alkalinity of the effluent of the laboratory-scale activated sludge process were 12 mg SCOD l^{-1}, 0.5 mg NH_4^+-N l^{-1}, 8.4 mg NO_3^--N l^{-1} and 32.7 mg (as $CaCO_3$) l^{-1}, which were quite comparable to the corresponding values of 11 mg SCOD l^{-1}, 1.6 mg NH_4^+-N l^{-1}, 8.9 mg NO_3^--N l^{-1} and 37.0 mg (as $CaCO_3$) l^{-1} of the effluent of the full-scale activated sludge process. The percentages of COD and nitrogen assimilated into the biomass and the COD and nitrogen constituents in the effluent were comparable with those of the full-scale process of Phase IV of Bedok WRP.

The comparison of the results of the laboratory experiment with the full-scale system performance illustrated that the scale-down principle and the 'ruling regimes' of the activated sludge process defined in this study were workable. Such laboratory-scale process experimentation provides a cost-effective tool in predicting the full-scale system performance, optimizing the existing process, and developing new processes.

Optimization of the BNR activated sludge process in warm climates. Under an aerobic SRT of 3.75 d and with the anoxic volume increased from 25% to 62.5% (corresponding to 6.25 d anoxic SRT) of the total tank volume and feed conditions of Phase IV of Bedok WRP, the laboratory-scale MLE process showed complete nitrification demonstrating the feasibility of significant reduction of the aerobic SRT and the volume adopted in the existing activated sludge processes.

Denitrification efficiency in the expanded anoxic reactors was 55.7% when the anoxic volumetric ratio was 62.5%. It was higher than 37.3%, the denitrification efficiency in the anoxic compartment of the full-scale activated sludge process of Bedok WRP where the anoxic volumetric ratio was 25%. The oxygen requirement reduction was 16% compared with 12%, the data of the full-scale system with 25% anoxic volumetric ratio. Also, increases in average pH and alkalinity in the aerobic reactor of 0.2 units and 10 mg as $CaCO_3$ l^{-1}, respectively, were recorded. These results demonstrated that denitrification in the existing WRPs could be further enhanced by expanding the anoxic zone with the volume increase realized through reduction of the aerobic zone.

Laboratory optimization studies illustrated that in a warm climate, nitrification could be achieved at an aerobic SRT of 3.75 d, and denitrification can be largely enhanced by enlarging the existing anoxic volumetric ratio from 25 to 50% or more of the total volume.

6.5 MODELING WITH ACTIVATED SLUDGE MODEL NO. 1 (ASM NO. 1)

A COD-based influent model, which enables the use of regular monitoring data as input, was developed and adopted. Primary calibration of the parameters of ASM No. 1 was made with a set of data from a laboratory-scale study. Parameter verification was undertaken with five different sets of diurnal data obtained from both laboratory-scale and full-scale investigations of two different activated sludge processes that described the performances of the activated sludge processes of Bedok, Kranji and Seletar WRPs.

COD-based influent model. A COD-based influent model developed based on the results of the sewage characterization in warm climates allows the use of COD, SCOD, TSS, VSS, TKN and NH_4^+-N analyses in the regular monitoring/sampling programme, together with the suggested and calculated conversion coefficients, as data for the influent model in modeling with ASM No. 1. The results of parameter identification indicated that the methods suggested for the determination of the conversion coefficients of the COD-based influent model were workable. This largely simplifies the characterization work and eases the application of modeling and simulation in warm climates.

Parameter identification. The kinetic parameters and stoichiometric coefficients of heterotrophs and autotrophic nitrifiers were first calibrated with the measured data obtained from the laboratory experiment of the MLE process of Phase IV of Bedok WRP. Only three parameters, i.e., $\hat{\mu}_A$, anoxic Y_{HD}, and b_H were regulated during the calibration, and the other parameters were taken either from the default values or the values after temperature correction.

The calibrated kinetic parameters and stoichiometric coefficients were verified separately with five sets of measured data from different activated sludge processes of both laboratory- and full-scale processes, including MLE, simultaneous nitrification and denitrification (SND), Bardenpho and step-feed activated sludge processes, of the three different WRPs under dynamic feed conditions. In these verifications, the general fits between the simulated and measured NH_4^+-N and NO_3^--N concentration profiles were satisfactory. No further calibration and verification of parameter values was required.

6.6 DESIGN OF THE BNR ACTIVATED SLUDGE PROCESS IN WARM CLIMATES

An aerobic SRT of about 4 days for nitrification can be adopted in the design of BNR activated sludge process in warm climates corresponding to a minimum aerobic SRT of about 2 days. This aerobic SRT is about half of that of the existing activated sludge process. An anoxic SRT of between 3 and 5 days is proposed for denitrification. It is longer than that of the conventional design because the concentration ranges of both the readily biodegradable COD and alkalinity of the sewage in Singapore are near the low boundary of the respective ranges of reported values. The final selection of anoxic SRT depends on the sewage characterization; the higher the total COD and readily biodegradable COD, the shorter the anoxic SRT, while the lower the total COD and readily biodegradable COD, the longer the anoxic SRT.

6.7 UPGRADING THE EXISTING BNR ACTIVATED SLUDGE PROCESS IN WARM CLIMATES

A MLE process with an anoxic and an aerobic volumetric ratio of 50% each is the recommended configuration proposed for the existing activated sludge process to be upgraded to. With respect to the existing activated sludge process, the current anoxic volume is to be expanded from 20-25% to 50% while the current aerobic volume is to be reduced from 75-80% to 50%. The total SRT of 10 d and the HRT of 6 h, which are similar to those of the existing process, can be retained. Both MLR and RAS ratios are 100% each but smaller ratios can be adopted during the low (base) flow period.

The simulations identified the advantages of expanding the anoxic volume, reducing the aerobic volume and adopting the recommended operational strategies as follows:

i   denitrification was enhanced by 29% with the increased anoxic volume;

ii   the daily average NH_4^+-N concentration was « 5 mg NH_4^+-N l^{-1} for an aerobic SRT of 5 d, even with a peak flow factor of 2.0;

iii the biomass concentrations in the reactors reduced during the peak flow period and increased during the base flow period. This biomass reduction could cause the effluent NH_4^+-N concentration to increase during the peak flow period. Adoption of a high RAS ratio during the peak flow period and a low RAS ratio during the base flow period could minimize the effluent NH_4^+-N concentration fluctuations; and

iv   the air supply to each compartment could be regulated according to the oxygen demand during the day and at night to save aeration energy further.

6.8 IMPLICATIONS FOR THE PROCESS DEVELOPMENT OF THE ACTIVATED SLUDGE PROCESS

The difference in scale between the laboratory- and full-scale systems is in the order of magnitude of 1,000,000 times, i.e. about 6 litres versus 8,700m^3. The magnitude difference is the largest ever reported in the scale-down of laboratory experiments and in scale-up modeling to study full-scale activated sludge processes. The success in parameter identification illustrated that: (i) the methodologies and approaches of the study were successful; (ii) the scale-down principles developed in the design of a laboratory experiment to investigate a full-scale activated sludge process were workable; and (iii) the approach to calibrate a few model parameters obtained from a laboratory experiment to predict the performance of a full-scale system was feasible.

Given the technology advancements to date, there are no principal difficulties in establishing a laboratory-scale system at a reasonable cost to simulate a full-scale activated sludge process under dynamic feed conditions, even with the incorporation of daily and seasonal temperature changes, non-ideal mixing and dead space in the full-scale process into the design. The outcomes of this study are encouraging and demonstrate the feasibility to adopt a laboratory experiment to study and model existing full-scale systems and design new processes. *From Flask to Field* appears to be a real possibility rather than purely a dream in activated sludge wastewater treatment research and development, and it could save a lot of time and resources.

Index

settled sewage 9--29

sewage denitrification potential 24

simulation

  laboratory 6, 7, 51--82

    activated sludge process 52--3

    Bedok Water Reclamation Plant 53--67

simultaneous nitrification and denitrification 42--3

site conditions 32--3

solid retention time 6

  minimum aerobic 117--19

soluble chemical oxygen demand 15--19, 27, 36, 39

soluble chemical oxygen demand removal 56

specific nitrification rate 37

SRT *see* solid retention time

stoichiometry 85

systematic approach 5

T

TKN *see* total Kjeldahl nitrogen

TN *see* total nitrogen

total Kjeldahl nitrogen 10, 17, 19, 35

total nitrogen 18

total phosphorus 17, 19, 35

total suspended solids 10, 15--19

TP *see* total phosphorus

TSS *see* total suspended solids

U

United States Environment Protection Agency 1

V

volatile fatty acids 10

volatile suspended solids 10

VSS *see* variable suspended solids

W

warm climates, optimization for 67--79

  carbonaceous matter removal 70

  denitrification 72--5

    activated sludge tanks 72--3

    final clarifier 74--5

  effluent quality 76--7

  experimental design 68--9

  mass balance and yield coefficients 77- -9

  COD balance 77--8

nitrogen balance 78--9

  nitrification 70--2

  pH 75

  shorter solids retention time 67--8

wastewater characterization 9--29

  advanced for modeling 10--11

  conventional 10

water reclamation plants 2--4

World Health Organization 1

Y

yield coefficient 45--8

  laboratory simulation 63--5

  optimization for warm climates 77--9

Printed in the United Kingdom
by Lightning Source UK Ltd.
129303UK00002B/83/P